Essential Basic Sciences for Orthopaedics

Essential Basic Sciences for Orthopaedics

Edited by

David Barrett, BS (Hons), FRCS
Consultant Orthopaedic Surgeon, Southampton General Hospital, Southampton

Foreword by

Professor George Bentley ChM, FRCS
Professor of Orthopaedic Surgery, Royal National Orthopaedic Hospital,
Stanmore, Middlesex

Butterworth-Heinemann Ltd
Linacre House, Jordan Hill, Oxford OX2 8DP

A member of the Reed Elsevier group

OXFORD LONDON BOSTON
MUNICH NEW DELHI SINGAPORE SYDNEY
TOKYO TORONTO WELLINGTON

First published 1994

British Library Cataloguing in Publication Data

Essential Basic Sciences for Orthopaedics
I. Barrett, D.S.
617.3

ISBN 0 7506 1639 3

Library of Congress Cataloguing in Publication Data

Essential basic sciences for orthopaedics/edited by D. Barrett.
p. cm.
Includes bibliographical references and index.
ISBN 0 7506 1639 3
1. Musculoskeletal system – Physiology. 2. Musculoskeletal system – Pathophysiology. 3. Orthopedics. 4. Surgical wound infections – Prevention. I. Barrett, D. S. (David Stuart)
[DNLM: 1. Musculoskeletal System – physiology. 2. Connective Tissue – physiology. 3. Musculoskeletal Diseases – physiopathology. 4. Musculoskeletal Diseases – surgery. 5. Surgical Wound Infection – prevention & control. WE 100 E78 1994]
RD732.E76
612.7 – dc20 93-32473
CIP

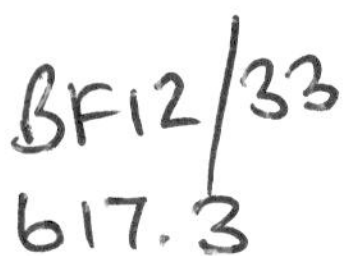

Printed and bound in Great Britain by Redwood Books, Trowbridge, Wiltshire

Contents

Foreword

In 1982 the Stanmore Professional Unit seminars began with the aim of filling an educational gap in the training programme which was heavily clinically orientated. Each lecturer, senior registrar and registrar was assigned a topic to prepare with a literature review and update of the subject which was illustrated and presented with slides. A text was prepared for all participants. With time and with the increasing enthusiasm of the trainees, these sessions have become increasingly valuable not only as a means of developing communication skills, but also as a stimulus to many of the young surgeons in training to carry out clinical and basic science research projects. Recently the material presented has been invaluable for candidates for the FRCS (Orth) diploma and the London MSc in Orthopaedic Surgery degree.

David Bárrett is to be congratulated on selecting and editing these basic science seminars which are particularly relevant for the present day trainee and practising orthopaedic surgeon alike.

George Bentley, ChM, FRCS
Professor of Orthopaedic Surgery

Introduction

Basic sciences and biomechanics form a significant part of the FRCS (Orth). Trainees preparing for the exam must currently study pure basic science textbooks where information may not be easily accessible or in a form relevant to the clinical practice of orthopaedics.

This book presents focused and appropriate information suitable for the FRCS (Orth). Topics which commonly form the basis of the viva or the written section have been prepared by senior registrars at the Royal National Orthopaedic Hospital who are studying for, or have already successfully completed, the examination. Each chapter is a review and discussion of an aspect of basic science in orthopaedics compiled from the primary sources and current literature. Sections are not intended to be lengthy and exhaustive works of reference but present basic science in a relevant and readable form as applied to orthopaedics.

The first chapters deal exclusively with basic biology topics, the latter section covers commonly occurring themes in the oral exam for which there is no recognized single textbook source.

Contributors

David S. Barrett BSc (Hons), FRCS
Southampton General Hospital, Southampton

Ian Dingwall FRACS (Orth)
Waikato Hospital, Hamilton, New Zealand

Simon T. Donell BSc, FRCS (Orth)
The Royal National Orthopaedic Hospital, Stanmore, Middlesex

Boyd Goldie BSc, FRCS
Whipps Cross Hospital, London

Phil L. Housden FRCS
Morriston Hospital, Swansea

George B. Irvine FRCS Ed (Orth)
Torbay Hospital, Torquay, Devon

Michael Kurer FRCS
North Middlesex Hospital, London

David Pring FRCS
Guernsey General Hospital, Channel Islands

Dishan Singh FRCS
The Royal National Orthopaedic Hospital, Stanmore, Middlesex

Allan N. Stirrat FRCS Ed
Sunderland District General Hospital

Roy S. Twyman FRCS (Orth)
The Middlesex Hospital, London

Chapter 1

Biology and mechanical properties of cartilage

Dishan Singh

Introduction

Cartilage consists of a relatively small number of cells and an abundant extracellular matrix. Macroscopically and under light microscopy, cartilage looks simple, homogeneous and inert when compared with the complexity and dynamic activity of other tissues (Caplan, 1984). Yet, the homogeneous appearance of cartilage hides a highly ordered complex structure (Fig. 1) that provides mechanical properties unmatched by any substitute.

Types of cartilage

Early investigators subdivided cartilage into three main types, which probably represent a continuum.

Hyaline cartilage is a glossy, opalescent material in which collagen fibrils are masked by the abundant basophilic ground substance of similar refractive index. It is found on most articular surfaces, costal cartilages, in the nasal septum and in some laryngeal and tracheal rings. It is the main component of the cartilaginous fetal skeleton and of the growth plates of children's bones.

Yellow fibrocartilage (synonym elastic cartilage) resembles hyaline cartilage under microscopy but on special staining, shows numerous bundles of branching elastic fibres in the cartilage matrix. It is present in the external ear, external auditory canal, epiglottis and parts of the laryngeal cartilage.

White fibrocartilage (synonym fibrocartilage) consists of alternating layers of cartilage containing matrix and chondrocytes and thick layers of dense collagen fibres orientated in the direction of functional stress. Fibrocartilage is present in the intervertebral disc (Buckwater, 1982), menisci of the knee, some articular cartilages and some connections of tendons to bone.

Elastic and fibrocartilage, having a predominance of fibres (Table 1.1), serve to resist tensile stresses while hyaline cartilage, with a higher proportion of ground substance, is able to resist compressive forces. Hyaline cartilage is the most commonly found type of cartilage in the human body and is the most typical, true cartilage, the other types being modifications of it. Articular cartilage covers the bones and synovial joints and is mostly of the hyaline variety, although it contains more collagen than other types of hyaline cartilage.

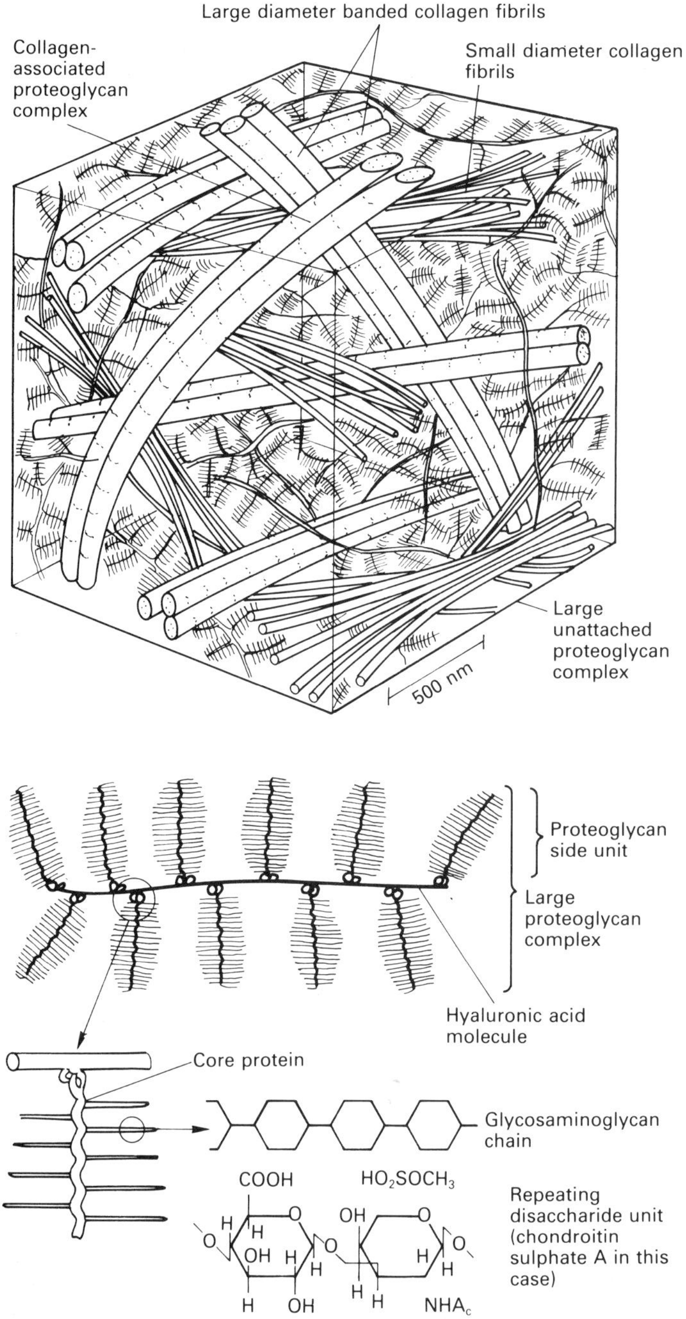

Figure 1.1 Structure of cartilage and proteoglycan. (From Gray's Anatomy, 37th edn, 1989, by kind permission of Churchill Livingstone, London.)

Table 1.1 Approximate composition of types of human cartilage

	Dry weight (%)		
	Collagen	*Proteoglycan*	*Elastin*
Hyaline	66	18	Nil
Fibrocartilage	78	2.4	Nil
Elastic	70	10	2.0

Histology of cartilage

The tissue consists of a sparse population of cells (chondrocytes < 5% by volume) housed in an abundant extracellular matrix. Although the cells and matrix are structurally separate, they are functionally very interdependent; chondrocyte activity is necessary for the synthesis and physiological degradation of matrix components, while the matrix, which consists of 60–80% water, plays an important part in maintaining the homeostasis of the chondrocyte's environment (Mitchell and Shepard, 1989). The physiological and mechanical properties of cartilage depend on the properties of the matrix; the main role of the chondrocyte is to manufacture and maintain the extracellular matrix.

The extracellular (or intercellular) matrix (see Fig. 1.1), far from being a homogeneous amorphous gel, has an organized macromolecular structure made up of fibrillar components (collagen and elastin) and a non fibrillar component termed ground substance (Fig. 1.2). The ground substance is a viscous gel made up of proteoglycans, glycoproteins, other proteins, lipids and minerals.

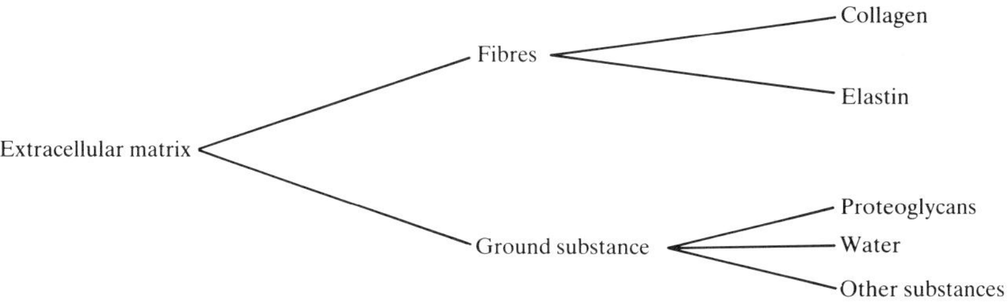

Figure 1.2 Composition of extracellular cartilage matrix

Fibres

Collagen

Collagen constitutes 40–70% of the dry weight of cartilage, its proportion decreasing with age. Under the electron microscope, collagen is seen to exist as fibres with typical cross-banding with a specificity of 67 nm (Fig. 1.3). Collagen fibres are flexible, but also have a high tensile strength with only a little elastic recoil and are vital to the mechanical properties of all connective tissues.

The basic unit of collagen is a polypeptide chain termed the procollagen chain. The procollagen consists of about 100 repeating amino acid triplets each with glycine at the third position, the other two varying in type but often being the amino acids hydroxyproline and hydroxylysine (substances rarely found outside collagen). Each alpha polypeptide chain forms a tight left-handed helix and then three helico-alpha chains wrap around each other in a right-handed helix to form a tropocollagen molecule, the configuration being made more stable by covalent and hydrogen bonds between the polypeptide molecules. The tropocollagen molecule (molecular weight 280 000 – 540 000) is 300 nm long and 1.4 nm thick.

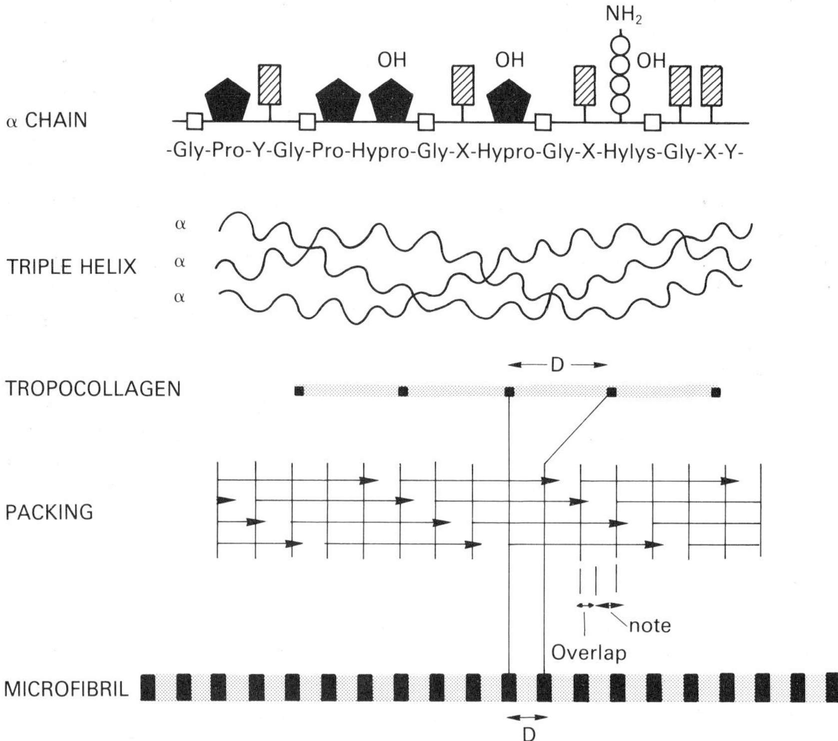

Figure 1.3 Organization of collagen microfibril. Top of figure shows part of amino acid sequence found in alpha chain. Glycine repeats as every third amino acid, proline and hydroxyproline appear frequently, and less common amino acids are designated by X or Y. Alpha chains spiral into tight left-handed helices, and then they wrap around each other to form triple helix or tropocollagen molecule. Pattern of amino acids in tropocollagen produces charged areas that stain darkly under certain conditions. Five of these charged areas, 68 nm apart, produce repeating period (D) of fibrillar collagen. (Adapted from information in: Gross, J. (1961) *Scientific American,* **204,** 120; Grant, M. and Prockop, D. (1972) *New England Journal of Medicine,* **286,** 194; and Eyre, D.R. (1980) *Science,* **207,** 1315.)

The tropocollagen molecules then aggregate into microfibrils (3.5 nm thick) which, in turn, aggregate into collagen fibrils (20–200 nm thick) which further aggregate into collagen fibres. The collagen fibres are again stabilized by covalent and hydrogen bonds and have a high tensile strength (15–30 kg/mm^2), weight for weight equivalent to steel. The precursor molecules of the collagen fibres in cartilage are synthesized and secreted by the chondrocytes, but fibrogenesis occurs extracellularly. The exact mechanism whereby fibrogenesis is controlled is not known.

Biochemical analysis and immunohistochemical detection (Williams *et al.*, 1989) has shown that there is a large family of chemically related polypeptides that make up collagen. The particular combination affects the structural properties of collagen, e.g. ability to form fibrils, diameter of fibrils, binding to adjacent structures and degree of hydration. Type I collagen is the most widely distributed in all connective tissues. Collagen of fibrocartilage is either exclusively type I (e.g. knee meniscus) or a mixture type I and type II (e.g. intervertebral disc). In hyaline cartilage most of the collagen is of the type II variety, but the natural repair mechanism of articular cartilage in joints is by the formation of fibrocartilage containing collagen type I predominantly (Stockwell, 1987). More recently (Gadher *et al*, 1990), collagen types IX, X and XI have been identified in articular hyaline cartilage. However, their exact significance is not known. Collagen type IX is causing much interest as it is believed to be covalently linked to the surface of type II fibrils forming a sheath around the type II fibril (Vaugham *et al.*, 1988) and to have projecting polysaccharide branches which form attachments with the proteoglycans. It may be relevant that type IX collagen is not degraded by collagenase.

Biomechanically, the spiral in the collagen fibril may be considered as a long, coiled spring. The collagen fibre in turn may be considered as an assemblage of springs connected together by a great number of less stiff spring like cross-links. Small stresses tend to uncoil the springs, whereas larger stresses tend to break and remake the interfibril cross links (Black, 1988).

Elastin

Elastin, a polypeptide in which about 60% of the amino acids are hydrophobic, possesses a spiral form but does not form large scale fibres. In its natural aqueous state, the elastin molecule organizes itself so that the hydrophobic amino acids move into close proximity in the interior of the globular macromolecule. A simplified molecular-kinetic theory of elastic deformation is that the elastic recoil is due to the tendency towards the maximal hydrophobic interaction after stretching. Elastin is able to sustain large deformations without rupturing and to revert spontaneously to its original condition when tension is released. In elastic cartilage elastin is mixed with collagen, which is more resistant to stretching.

The ground substance

Proteoglycans

Proteoglycans are large hydrophilic molecules consisting of a linear core protein to which are attached 50–100 side chains of the glycosaminoglycan chondroitin sulphate and keratan sulphate (see Fig. 1.1). Glycosaminoglycans (formerly called mucopolysaccharides) refer to long unbranched carbohydrate chains made up largely of repeating disaccharide units which have high negative charges due to attached carboxyl and sulphate groups. These high fixed negative charges attract large numbers of positively charged ions, and the resultant high osmolality accounts for the retention of water in cartilage.

Proteoglycans exist in two forms: monomer and aggregates. Under normal physiological considerations, the proteoglycan molecules (referred to as proteoglycan monomers) can interact specifically with hyaluronic acid to form enormous multi-molecular aggregates.

Hyaluronic acid + Proteoglycan monomer   Proteoglycan aggregate

The structure of a proteoglycan aggregate is stabilized by interaction with a glycoprotein known as link protein. Much smaller aggregates are formed in the absence of link proteins. Various subtypes of proteoglycans and link proteins have been described (Carney and Muir, 1988; Larsson, Aspden and Hukins, 1989).

The resultant proteoglycan aggregate resembles a bottle brush (see Fig. 1.1). The long axis of the bottle brush analogy is a stretched out molecule of hyaluronic acid. Each of the proteoglycan molecules that extend laterally like a bristle from its axis is itself made up of a long central core protein molecule to which is attached a number of proximal side chains of keratan sulphate and an even greater number of distal side chains of chondroitin sulphate. Mutual electrostatic repulsion between the negative charges on the glycosaminoglycan molecules expands the proteoglycan aggregate to its maximal volume and provides innumerable domains to harbour water molecules and ions. This mechanism provides cartilage matrix with an intrinsic mechanism of resilience: this is because an applied compressive force displaces water from the negatively-charged domains on the proteoglycan chains, bringing the negatively-charged domains into closer proximity whereupon mutual repulsion opposes the compressive force that is responsible for such displacement.

The proteoglycan–hyaluronide aggregates are very large (molecular weight about 50×10^6) and are one of the largest proteins made by living organisms. The proteoglycan–hyaluronide aggregates are trapped in the meshwork of collagen fibrils due to their mere size; also specific chemical interactions between the proteoglycan and the collagen fibres may be present. In situ, in normal cartilage, the proteoglycan aggregates are compressed to about one fifth of their potential volume and have a constant tendency to swell. They are prevented from swelling to their maximum volume by their topographical entanglement within the fibrous collagen meshwork and, in turn, the collagen fibrils and fibres are put under stretch, thereby increasing their tensile properties. Thus, the biomechanical properties of cartilage are dependent on this interaction between the collagen fibres and the proteoglycan aggregates (Broom, 1984).

Relatively little is known about how the assembly of proteoglycans into aggregates outside the cell is controlled and even less about how these aggregates are directed to appropriate sights within an existing matrix.

Other substances in the ground substance

Glycoproteins and other proteins have been described as playing important structural roles (Williams *et al.*, 1989). Cell adhesion proteins attach chondrocytes to matrix components, e.g. chondronectin and anchorin. Large numbers of enzymes are present, including collagenase and elastase. The exact significance of the presence of lipids, with their concentration increasing with age, is not clear. Minerals are present in the matrix and increased amounts of apatite crystals are found in osteoarthritic cartilage (Ali and Griffiths, 1983).

Chondrocytes

Cartilage is almost unique among tissues in that it contains only one type of cell and is completely free from any wandering cells found in other tissues. On light microscopy, chondrocytes lie in lacunae but, with electron microscopy, it is believed that the chondrocytes fully occupy the cavity in the matrix. The internal structure of chondrocytes is typical of cells active in making and secreting proteins, i.e. round nucleus, prominent nucleoli, granular endoplasmic reticulum and Golgi complexes. The chondrocytes have many projecting processes and the tips of these processes may become detached to form vesicles in the matrix. It is thought that these vesicles rich in alkaline phosphatase may be involved in the calcification of growth plates and in the development of osteoarthritis (Ali and Griffiths, 1983).

It was previously thought that chondrocytes in adult cartilage cannot undergo mitotic division; however, cell division has been demonstrated (Rothwell and Bentley, 1973).

It is now well established that all of the major components of a matrix are synthesized and secreted by chondrocytes. The same cell can synthesize the precursors of collagens and proteoglycans. During skeletal growth chondrocytes proliferate rapidly and synthesize large volumes of matrix. However, with maturation these processes slow down and the cell density decreases. The understanding of local cell control is poor. Systemic factors, hormonal control, concentration of hyaluronic acid, pressure, injury and drugs may affect the activity of a cell. Equally important, once the cells synthesize the molecules, what directs the organization of macromolecules in the matrix? Perhaps the cells synthesize appropriate types and amounts of molecules that diffuse through matrix and assemble spontaneously, or perhaps they synthesize different amounts and types of small molecules such as link proteins that help influence matrix organization. The chondrocytes also secrete metabolic enzymes which play a role in the slow turn-over of the matrix components. Just as the properties of cartilage depend on its extracellular matrix, the existence and maintenance of a matrix depends on the chondrocytes.

Articular cartilage

Articular cartilage is crucial to the normal function of synovial joints since its presence enables the articulating bones to transmit high loads while maintaining contact stresses at an appreciably low level, and to move on each other with little functional resistance. Articular cartilage shares its basic composition and structure with other hyaline cartilages, but to perform its specialized role as the surface of the synovial joint it has developed a complex internal organization (Buckwater, 1983).

Zones

Articular cartilage varies from 1 to 7mm in thickness. Morphological and biochemical examinations demonstrate four layers or zones proceeding from the articular surface to the subchondral bone (Fig. 1.4). The differences among zones reflect transition from resistance to prominent shear force at the joint surface to more compressive force deeper in the cartilage.

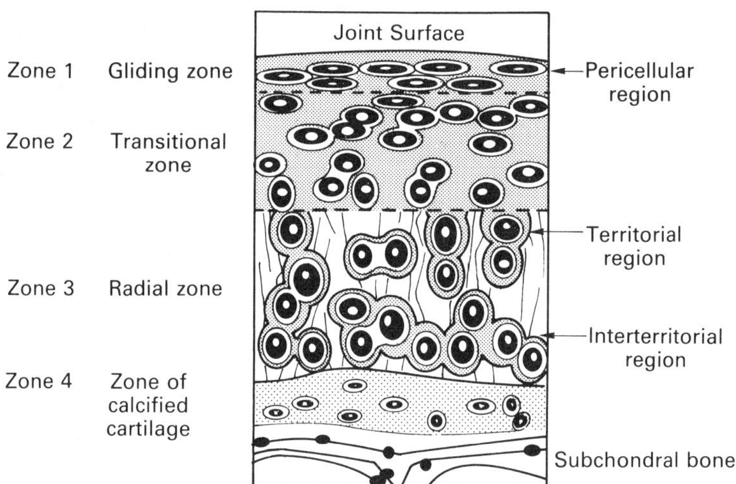

Figure 1.4 Diagram of articular cartilage showing four layers or zones progressing from articular surface to subchondral bone. Cells and matrix differ from one zone to next. Based on collagen fibril diameter and orientation, it may be possible to identify three regions within zone: pericellular region, territorial region, and interterritorial region. (From Buckwater, 1983, by kind permission of C.V. Mosby Co., St Louis.)

Zone I (superficial, tangential, gliding) has numerous discoid cells with their axes parallel to the surface and has abundant collagen fibrils arranged tangentially to resist tensile forces. Proteoglycan concentration is the lowest of all zones.

Zone II (intermediate, transitional, middle) and *zone III* (deep, radiate) have equally spaced spheroidal cells, a combination of random collagen fibrils forming a meshwork and high proteoglycan aggregate concentrations. Zones II and III therefore form the bulk of the cartilage thickness and provide most of the resistance to static compression loading. Zones I, II and III merge imperceptibly while zones III and IV meet at the weakly basophilic line designated the tide mark.

Zone IV (calcified) lies deepest, adjacent to the subchondral bone. It has few chondrocytes and most of the collagen fibres are arranged perpendicular to the subchondral bone. The matrix is impregnated with hydroxypatite, bonding the cartilage to the irregular surface of the subchondral bone.

The *surface* of cartilage is believed by some not to be perfectly smooth but to have the structure resembling that of a golf ball and it is thought that these pits play an important role in lubrication (reviewed by Bentley, 1987). The *marginal transitional zone* is at the periphery of the articular area of a synovial joint where articular cartilage, synovial membrane, joint capsule and bone merge. It is in this area that osteophytes, which are mainly made of fibrocartilage, arise.

Mechanism of load carriage

The arcades of collagen fibrils form a meshwork; the proteoglycan aggregates enmeshed in the collagen fibrils are not fully hydrated and have a tendency to swell. Aspden and Hukins (1989) have suggested that the mechanical analogy is that of a pressure vessel in which the internal swelling pressure is balanced by the tensile stress in the collagen fibril. Such a system is able to tolerate some damage to the surface zone because there are enough fibrils with the required orientation to withstand the internal pressure of the proteoglycan gel, e.g. laceration of the surface does not induce osteoarthritis (Mankin, 1982). The system will only become unstable when there are insufficient fibrils with the required pattern of orientation.

When a load is applied, water is driven away from the loaded region, thereby increasing the osmotic pressure or swelling pressure and an equilibrium will be reached when the load is balanced by the increase in swelling pressure of the concentrated ground substance. On load removal, water will be reimbibed until the proteoglycans occupy the maximum volume allowed by the collagen fibres, restoring the original equilibrium. Disruption of collagen fibre meshwork or degradation of the proteoglycans will therefore result in softer, less resilient cartilage.

If the load is removed after only a short time (quick loading), the articular cartilage returns rapidly to its original unloaded state. Under a prolonged load (slow loading), the cartilage will gradually decrease in volume due to a net loss of water; complete recovery will take a finite time. It has been established (Salter and Field, 1960) that a prolonged pressure applied experimentally to a joint causes degeneration of the cartilage together with damage to the chondrocytes. Under cyclical loading, the behaviour of the tissue will depend on the load, the frequency of loading and the number of load cycles.

Nutrition

The nutrition of articular cartilage has attracted much attention. Cartilage has no blood vessels and the proteoglycans and collagen content of a matrix regulate the passage of solids into and through the cartilage. Uncharged small molecules such as glucose and oxygen diffuse comparatively easily, whereas large molecules such as immunoglobulins are almost completely excluded. In the adult, most of the nutrients are obtained from the synovial fluid bathing the articular surface. Synovial fluid is essentially a dialysate of plasma with hyaluronic acid added. The role of diffusion from the subchondral marrow spaces across calcified cartilage in normal adult articulate cartilage is debated. The survival of cartilage in loose bodies and in avascular necrosis all suggest that synovial fluid is adequate to sustain articular cartilage.

It is possible that diffusion is made more rapid by the pumping action of synovial fluid induced by synovial joint movement (Freeman, 1979). Thus, the expression of fluid from the matrix upon load carriage, and the subsequent reimbibition of synovial fluid upon load removal, may contribute to the nutrition of the chondrocyte and may explain the degeneration of cartilage that occurs with rest.

Osteoarthritis

Osteoarthritis is a clinical syndrome of pain, stiffness and deformity with articular cartilage damage. The typical end-stage radiological changes include joint space narrowing (due to cartilage loss), subarticular sclerosis, cysts and osteophyte formation (due to increased activity of the underlying bone).

Early osteoarthritis is characterized by fibrillation of the superficial surface, loss of proteoglycans and collagen, and increase in water content (Bentley, 1985, 1987). The basal calcified layer expands, and thickening of the subchondral bone occurs. The arthritic joints become hypervascularized, but the venous outflow is delayed, resulting in a rise in the intraosseous pressure. Cartilaginous debris in the synovial fluid may lead to inflammation and subsequent fibrosis of the synovial membrane and joint capsule. Osteophyte formation involves de novo generation of fibrocartilage from the tissues at the transitional zone at the margin of the joint surface. Part of the osteophyte is ossific and radiopaque, but part is cartilaginous and not visible on X-ray films.

Osteoarthritis is an end-stage disease which is almost certainly not a single disease entity. The exact time sequence of events is debatable since the order in which the cells, ground substance matrix and collagen become altered is not at present fully known. It has been suggested (Kempson, Freeman and Swanson, 1968; Freeman, 1979) that osteoarthritis may be secondary to stress fracture of the collagen fibrils. A resultant swelling of a cartilage is due to distension of the proteoglycan aggregates; further damage causes stress fractures of more collagen fibrils and proteoglycan and collagen escape (Roberts *et al.*, 1986). Neighbouring chondrocytes divide and become metabolically more active in an attempt to repair the damage but fail to provide the normal macroarchitecture of articular cartilage.

Summary

Much work has been done in the last 30 years on the biomechanical structure of cartilage, but much remains to be understood. We no longer view cartilage as inert, homogeneous gristle but as a specialized tissue with a specialized macromolecular architecture in which the proteoglycan osmotic pressure maintains the arcades of collagen fibrils in a permanent state of turgor. It is this unique complex macroarchitecture that makes cartilage repair so complex.

References

Ali, S.Y. and Griffiths, S. (1983) Formation of calcium phosphate crystals in normal and osteoarthritic cartilage. *Annals of the Rheumatic Diseases,* **42**(Suppl 1), 45–48.

Aspden, R.M. and Hukins D.W.L. (1989) Stress in collagen fibrils of articular cartilage calculated from their measured orientations. *Matrix*, **9**, 486–488.

Bentley, G. (1985) Articular changes in chondromalacia patellae. *Journal of Bone and Joint Surgery,* **66B**, 769–774.

Bentley, G. (1987) Pathogenesis of osteoarthritis. In: (Hughes, S.P.K., Benson, M.K.D'.A. and Colton C.L., eds). *Orthopaedics, the Principles and Practice of Musculoskeletal Surgery.* Churchill Livingstone, London.

Black, J. (1988) *Orthopaedic Biomaterials in Research and Practice.* Churchill Livingstone, New York.

Broom, N.D.(1984) Further insights into the structural principles governing the function of articular cartilage. *Journal of Anatomy,* **139**, 275–294.

Buckwater, J.A. (1982) Fine structural studies of human intervertebral disc. *American Academy of Orthopedic Surgeons: Symposium on Idiopathic Low Back Pain.* C.V. Mosby Company, St Louis.

Buckwater, J.A. (1983) Articular Cartilage. *American Academy of Orthopedic Surgeons' Instructional Course Lectures*, **32**. C.V. Mosby Company, Toronto.

Caplan, A.I. (1984) Cartilage. *Scientific American*, **251**, 82–90.

Carney, S.L. and Muir, H. (1988) The structure and function of cartilage proteoglycans. *Physiological Reviews*, **68**, 858–910.

Freeman, M.A.R. (ed). Adult Articular Cartilage. Pitman Medical Publishing Co. Ltd. Kent, England, 1979.

Gadher, S.J., Eyre, D.R., Wotton, S.F., Schmid, T.M. and Woolley, D.E. (1990) Degradation of cartilage collagens type II, IX, X and XI by enzymes derived from human articular chondrocytes. *Matrix,* **10**, 154–163.

Kempson, G.E., Freeman, M.A.R. and Swanson, S.A.V. (1968) Tensile properties of articular cartilage. *Nature*, **220**, 1127–1128.

Larsson, T. and Aspden, R.M. and Heinegard D. (1989) Large cartilage proteoglycan influences the biosynthesis of macromolecules by isolated chondrocytes. *Matrix*, **9**, 343–352.

Mankin, H.J. (1982) The reaction of articular cartilage to mechanical injury. *Journal of Bone and Joint Surgery,* **64A**, 460–466.

Mitchell, N. and Shepard, N. (1989) The deleterious effects of drying on articular cartilage. *Journal of Bone and Joint Surgery,* **71A**, 89–95.

Roberts, S., Weightman, B., Urban, J. and Chappell, D. (1986) Mechanical and biochemical properties of human articular cartilage in osteoarthritic femoral heads and in autopsy specimens. *Journal of Bone and Joint Surgery,* **68B**, 278–288.

Rothwell, A.G. and Bentley, G. (1973) Chondrocyte multiplication in osteoarthritic articular cartilage. *Journal of Bone and Joint Surgery,* **53B**, 588–594.

Salter, R.B. and Field, P. (1960) The effect of continuous pressure on articular cartilage. *Journal of Bone and Joint Surgery,* **42A**, 31–38.

Stockwell, R.A. (1987) Cartilage. In: (Hughes, S.P.F, Benson, M.K.D'.A., Colton, C.L., eds). *Orthopaedics, the Principles and Practice of Musculoskeletal Surgery.* Churchill Livingstone, London.

Vaugham, L. *et al*. (1988) Periodic Distribution of collagen type IX along cartilage fibrils. *Journal of Cell Biology,* **106**, 991–997.

Williams, P.L., Warwick, W., Dyson, M. and Bannister, L.H. (1989) *Gray's Anatomy*, 37th edn. Churchill Livingstone, London.

Chapter 2

Articular cartilage damage and repair

Simon T. Donell

Introduction

Articular cartilage is a highly specialized tissue, and is relatively isolated. It has neither blood, lymphatic, nor nerve supply. This arises because it exists in a highly mechanical environment with loads varying from zero to many times body weight passing through it. In these circumstances it would be rendered ischaemic for much of the time during activity, and this therefore precludes it from relying on a blood supply. It therefore follows that it does not require lymphatics. Nutrients and waste products are diffused through cartilage by the movement of synovial fluid. The movement comes about by the changing loads and pressure gradients that occur during physiological activity. Recent in vitro work on human femoral heads by O'Hara, Urban and Maroudas (1990), however, has suggested that the pumping effect is only important for large molecules, whereas small molecules such as oxygen and glucose can dissolve freely even when cyclical loading occurs.

The mechanical extremes that articular cartilage can experience have the potential to cause significant damage, and therefore the need for repair may arise. However, the repair mechanisms of articular cartilage are poor. Despite this the majority of synovial joints in the majority of people work perfectly satisfactorily for all of their lives. Articular cartilage must therefore exist in a balanced environment, that is balanced biochemically and physically. This chapter reviews the ways this environment can be damaged and the ways in which it may be repaired.

Mechanics of articular cartilage (Mow, Proctor and Kelly, 1989)

Function

Articular cartilage has two main functions: to distribute joint loads and so reduce the stresses experienced, and to allow movement between opposing surfaces with the minimum of friction and wear. It is important to remember that articular cartilage does not exist in complete anatomical isolation and that the properties of synovial fluid and the functions of the synovial membrane are clearly important in cartilage homeostasis, as well as the mechanical effects of the subchondral bone.

Biomechanical properties

Articular cartilage is a viscoelastic substance that is biphasic. The fluid phase is composed of water with dissolved inorganic salts. The solid phase is made up of an organic solid matrix, i.e. collagen and proteoglycans. It can be likened to a water-soaked sponge. The water concentration differs in the various layers of cartilage, being 80% at the surface and falling to 65% in the deep zone. Seventy per cent of the water is intermolecular and is available to move when a load or pressure gradient is applied.

Articular cartilage is also anisotropic. This means that it has different mechanical properties depending on the direction it is loaded. This is due to the relationships of the collagen fibre arrangements, their cross-linking, and variations in the collagen–proteoglycans interactions.

Creep and stress relaxation

If a constant load is applied to articular cartilage it will initially deform rapidly and then show slow progressive increasing deformation, known as creep, until a steady state is reached. If constant deformation is applied there is a high initial stress followed by a slow progressive decreasing stress to the level that is required to maintain the deformation. This is known as stress relaxation. 'Slow' means that it is time-dependent. Both creep and stress relaxation occur through interstitial fluid movement and macromolecular movement. Excessive stress is difficult to achieve at physiological loads because of the effects of stress relaxation. Compression tends to cause movement of fluid, whereas shear tends to cause movement of macromolecules.

Permeability

Articular cartilage is porous and freely permeable. However, under high loads the movement of water is hindered by the effect of the frictional drag of the macromolecules. This causes a decrease in the flow and therefore stiffens the tissue, allowing greater resistance to the higher load.

Tensile effects

Tension alters the molecular structure of articular cartilage, the organization of collagen fibres, and the collagen fibre cross-links. They are, in effect, pulled apart. The tissue will become more swollen as permeability is increased and therefore fluid moves in. This leads to a decrease in the compressive stiffness.

Lubrication

Despite an enormous range of differing loading conditions articular cartilage suffers from little wear. This is due to a very sophisticated lubrication system that has both adsorbed lubricant on the articular cartilage surface (boundary lubrication), and lubricant in a film between the surfaces of the articulating cartilage (fluid film lubrication). The lubricant in boundary lubrication is a glycoprotein called lubricin. The type of fluid film lubrication thought to occur in synovial joints is termed elastohydrodynamic. The sliding action of the articular surfaces forms a squeeze film and the pressure generated deforms the cartilage to increase the bearing contact area which, in turn, decreases the ability of the film to be dissipated. It therefore allows high load-bearing. Furthermore, under load, fluid is forced out of the cartilage adding lubricant to the system during movement. Under extreme loading conditions, such as prolonged standing after impact, surface-to-surface contact may occur, but the surfaces are still protected by the layer of lubricin. The above comments are a simplification of the known properties of synovial lubrication, but the essential feature is that there are a number of different lubricating mechanisms for different loading conditions.

Wear

Wear is the removal of material from solid surfaces by mechanical action. It can be interfacial, due to the action of the bearing surfaces, or fatigue, due to bearing deformation under load.

Interfacial wear can be adhesive (the surfaces come into contact and fragment) or abrasive (when a soft material is scraped by a harder one). Interfacial wear is controlled by adequate lubrication. If the mechanical properties of the cartilage are altered, e.g. by altered permeability, leading to softening, lubricant may leak away and allow interfacial wear. This appears to occur in degenerative arthritis.

Fatigue wear occurs when repetitive stressing causes an accumulation of microscopic damage. This can be due to either high stress and low cycle loading, or low stress and high cycle loading. In the latter damage can occur when the loads are below the material's ultimate tensile strength, and even when well lubricated. Three mechanisms of damage have been proposed:

1 Disruption of the collagen–proteoglycan matrix, i.e. tensile failure of the collagen fibre network, a state that has also been noted with increasing age and in some diseases.
2 Leeching or washing out of proteoglycan by repeated large movements of interstitial fluid in the superficial cartilage layer causing increased permeability and decreased stiffness.
3 Rapid repeated high loading, where there is no time for stress relaxation, and so resulting in collagen–proteoglycan matrix damage.

The macroscopic effects of this damage can cause vertical splits (fibrillation) which may penetrate through the whole depth of the cartilage. Erosions may occur due to smooth surface destructive thinning. The critical point is that once the collagen–proteoglycan matrix is disrupted by the mechanisms described above, these processes can accelerate the rate of interfacial and fatigue wear. The mechanical destruction of the collagen-proteoglycan matrix is summarized in Figure 2.1.

Collagen-proteoglycan matrix disruption can occur either at a mechanical level, e.g. by high stress from joint incongruity or ligament rupture, or at a biochemical level, e.g. by disordered collagen metabolism, or proteolytic enzyme degradation. It should also be appreciated that traumatic effects will initially cause localized injury, whereas at a biochemical level articular damage is likely to be diffuse.

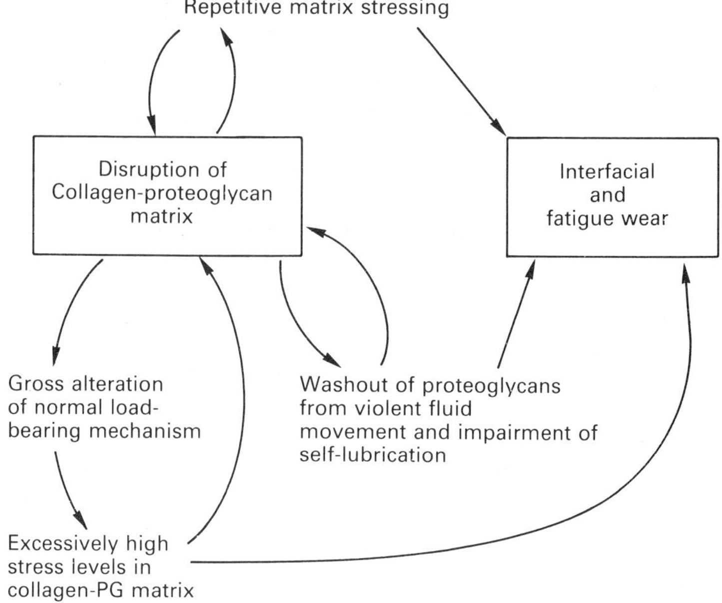

Figure 2.1 Mechanical effects of collagen-proteoglycan matrix disruption.

Chondrocyte function

Cartilage is an unusual tissue in that it contains only one cell, the chondrocyte. Migratory cells do not appear in normal articular cartilage. In mature adult cartilage the number of chondrocytes is sparse (less than 10% of the tissue volume). This is reflected in the low metabolic rate of articular cartilage. Higher rates cannot be achieved as nutrients and degradation products pass in and out only by diffusion, even though this may be enhanced by the pumping effect produced by cyclical loading. Chondrocytes manufacture, secrete and maintain the organic matrix, producing both the collagen and the proteoglycans. Maintenance of the organic matrix implies an ability to remove and replace the molecules involved. The mechanisms employed in arranging the matrix are unknown, however, the recent production of a chondrocyte line that does not degrade with time may begin to yield information in this area (Block *et al.*, 1991).

Biochemical aspects of cartilage destruction

As stated earlier the key to cartilage destruction is degradation of the collagen–proteoglycan matrix. The mechanical aspects of this have already been discussed. Direct effects on the organic matrix can occur from inborn errors of collagen metabolism (Mayne, 1989; Hull and Pope, 1989), the destructive effects of free radicals (Blake *et al.*, 1989), or destructive molecules from synoviocytes, chondrocytes or migratory cells such as polymorpholeucocytes. Management of gene defects is outside the remit of this chapter and halting the destructive effects on articular cartilage of recurrent haemarthrosis or septic arthritis is obvious. This chapter therefore concentrates on the mechanisms involving the chondrocytes and synoviocytes.

The biochemical pathways involved in cartilage destruction are the subject of much research at the present time. Figure 2.2 shows a possible scheme. Both the chondrocytes and synoviocytes produce the cytokines tumour necrosis factor-α (TNF-α) and interleukin 1 (IL1) which in turn cause these cells to synthesize metalloproteases such as collagenase and stromelysin (the latter detected via a caseinase reaction). TNF-α and IL1 have a complex interaction and are not purely synergistic or antagonistic. TNF-α seems to be the more potent stimulator of metalloprotease production. The metalloproteases have direct effects on the collagen–proteoglycan matrix causing it to breakdown, as indeed may IL1 (Pelletier and Martel-Pelletier, 1989). TNF-α also stimulates synoviocytes to synthesise prostaglandins, in particular prostaglandin E2 (PGE2) (Bunning and Russell, 1989). This causes plasma to be extravasated into the synovium leading to synovitis, but inhibits lymphocyte proliferation. PGE2 also has an inhibitory effect on IL1 thus reducing its effect on metalloprotease production and cartilage matrix breakdown. TNF-α is inhibited in its turn by γ-interferon (γ-IFN) which is produced by T lymphocytes in response to interleukin 2 (IL2). The latter is found in large quantities in rheumatoid arthritis (Pettipher *et al.*, 1989; Vignon *et al.*, 1990).

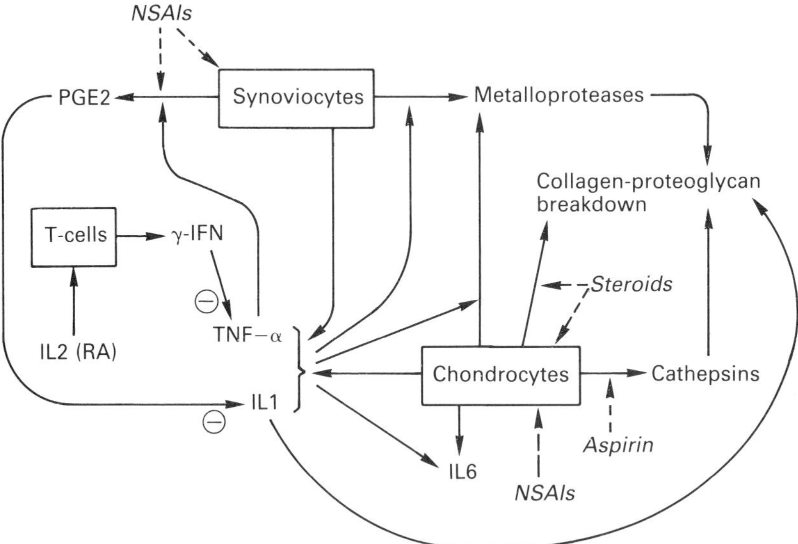

Figure 2.2 Possible biochemical pathways involved in cartilage destruction. IL1, IL2, IL6 = interleukins 1, 2 and 6; TNF −α = tumour necrosis factor −α; γ−IFN = γ−interferon; NSAIs = non-steroidal anti-inflammatory drugs; RA = rheumatoid arthritis

Chondrocytes are also known to produce cathepsins when damaged. These originate from lysosomes and have a direct action on proteoglycan breakdown (Bentley, Leslie and Fischer, 1981).

Finally chondrocytes produce interleukin 6 (IL6) in large quantities under the stimulus of TNF-α and IL1. Its direct action is unknown at present but it may have an inhibitory effect on metalloprotease synthesis (Shinmei *et al.*, 1990). Interestingly both IL1 and IL6 are known to stimulate liver cells to produce C-reactive protein (Yamada *et al.*, 1990). A summary of the cells involved and the reported effects of the cytokines found in cartilage degradation is shown in Table 2.1.

Table 2.1 Reported sites of production and effects of cytokines involved in cartilage destruction

Cytokine	*Chondrocyte*	*Synoviocyte*	*T cell*	*Effect*
TNF-α	+	+		↑ metalloproteinases ↑ prostaglandin
IL-1	+	+		↑ bone resorption
γ-IFN			+	synoviocyte proliferation ↓ bone resorption by TNF-α ↓ cartilage resorption by TNF-α
IL-6	++			?

IL-1 and IL-6 = interleukins 1 and 6; TNF-α = tumour necrosis factor-α; γ-IFN = γ-interferon.

Therefore, there are two cells directly involved in cartilage breakdown, the synoviocyte and the chondrocyte. Early synoviocyte-initiated damage will cause diffuse superficial cartilage destruction, whereas early chondrocyte-initiated damage can be local and involve the deeper layers, as would be expected after trauma, or more diffuse, if production of abnormal collagen is involved. Once the stage of collagen–proteoglycan disruption has occurred to any significant extent, then the destructive processes on the cartilage will be both mechanical and biochemical. Disintegrating cartilage fragments have a strong inflammatory effect on synovium, and areas of abnormal cartilage will be subject to adverse mechanical loading. A cycle of progressive destruction is then in place, as is typically observed in osteoarthritis.

Repair mechanisms

Most tissues respond to injury in broadly the same way. Initially there is a phase of necrosis, when cell death occurs. This depends on the size, extent, and duration of the insult, as well as the dependence of the tissues on its local blood supply and the availability of a collateral circulation. The second phase is inflammation. This will occur in infections and immune challenges, as well as trauma. Vascular channels dilate and increase blood flow locally. An extracellular exudate forms, rich in protein and migratory cells. This results in a fibrin clot which 'glues' the wound together. The final phase is repair. The inflammatory cells change to fibroblasts, blood vessels invade the fibrin clot to form granulation tissue, and this, in turn, changes into scar tissue. Some tissues, such as bone, can replicate identical tissue to replace damaged tissue, thus instead of scar, callus is formed. However, the basic repair process is the same (Mankin, 1982).

With cartilage there is no blood supply, so after necrosis of chondrocytes there is no phase of inflammation, and therefore no phase of repair into scar tissue. In fact, disruption of the collagen–proteoglycan matrix in itself is unlikely to lead to diffuse chondrocyte death. These circumstances seem to occur in chondromalacia patellae (CMP) (Ohno *et al.*, 1988) and some repair has been observed with the chondrocytes producing type II collagen (Bentley, 1985). This is a true type of primary repair. It is, however, slow and, if defects are large, incomplete. It is termed *intrinsic*. If the injury should involve exposure of the subchondral blood supply, then the classical type of repair is then possible, as there is now a blood supply. Fibroblasts can invade, and a fibrous scar formed. Under the influence of loading, metaplasia into fibrocartilage then occurs, with high levels of type I collagen. This is termed *extrinsic* repair.

Experimental injuries

Traumatic

1. Superficial laceration

In animal studies, mainly on rabbits, it has been found that within 24 hours of a superficial laceration, living chondrocytes adjacent to the wound replicate and start producing matrix. At the same time catabolic processes proceed with the production of degradative enzymes, presumably removing damaged matrix and any dead chondrocytes. However, this activity ceases after 2 weeks (DePalma, McKeever and Subin, 1966). Following this the laceration remains unchanged and does not proceed to osteoarthritis. This finding not only occurs with single superficial lacerations but also when scarifications with multiple linear cuts is made (Meachim, 1963). Tangential slices are found to have the same response (Ghadially *et al.*, 1977). At 6 months only minor remodelling occurs at the margins of the wound, and at 2 years the cartilage surface is, macroscopically, virtually the same as immediately after injury.

Almost all natural, as opposed to experimental or surgical injuries do not occur in isolation but are associated with damage to soft tissues which result in an haemarthrosis. Yet the blood does not clot and stabilize in the defect to form scar tissue. It is thought that proteoglycans inhibit clot formation since the proteoglycanolytic enzyme, papain, given in a low dose intra-articularly to rabbits allows clots to form (Farkas *et al.*, 1977).

A more recent study on mature rabbits repeated earlier work and showed no healing with a superficial laceration. However, if fibrin was added at the time of laceration, collagen bridging and healing was noted after 12 weeks (Kaplyoni *et al.*, 1988). It was argued that the repair appeared to be of the intrinsic type with primary healing by chondrocytes. This work has not been repeated to confirm the findings.

2. Ostechondral defect

If, however, the injury penetrates the subchondral endplate into bone, then a haematoma is rapidly formed enriched with fibrin and containing white blood cells and marrow cells. Fibroblasts invade and granulation tissue forms. Bone is formed at the base of the lesion, but it only fills to level of the subchondral bone plate. Hyaline cartilage at the edge of the wound shows a burst of synthetic activity, but this is short lived. The fibrous tissue undergoes metaplasia to form chondroid tissue, which eventually becomes fibrocartilage. There is no superficial tangential layer so, with time, this fibrillates, but it can remain stable for many years without progressing to osteoarthritis.

The size of the injury is important. In work on horses, Convery, Akeson and Keown, (1972) found that a defect less than 3 mm wide could be repaired in 3 months. If it was greater than 9 mm wide complete repair was never achieved. The repair tissue was a mixture of fibrous tissue, fibrocartilage and hyaline-like cartilage. The ultimate tissue contained 80% type II collagen and 20% type I. This is not pure hyaline cartilage, and eventually will undergo degeneration. Interestingly, in a study on dogs, Nelson *et al.* (1988) found no increase in stress concentration around 6 mm defects in the femoral condyles, although the defect was of poor mechanical quality.

3. Blunt trauma

Cyclical loading at different stresses, and for different times has shown that damage occurs at high strain rates and over long periods (Radin *et al.*, 1970; Radin *et al.*, 1978). The reasons for this have been described earlier.

4. Cavitation

Cavitation is the effect of bubble activity in fluids. The vaporous form of cavitation is harmful because it involves both explosive formation and collapse of the bubble into microbubbles, with wall velocities approaching the speed of sound. This can cause severe damage to surrounding structures. It is thought to be the source of the cracking sound in human metacarpophalangeal joints on distraction, and is distinguishable from other cracks and clicks in that it cannot be repeated until synovial fluid tension returns by absorption of the microbubbles formed. This takes about 20 minutes. Watson, Kernohan and Mollan (1989) induced cavitation in fresh cow knee joints using ultrasound. The effect was to produce a series of craters and pits on the articular surface. Comparison with human osteoarthritic cartilage showed remarkable similarities. It was pointed out that the normal joint is under subatmospheric pressure and that this could promote microbubble formation, which, as stated above, is protective against cavitation. They concluded that this was an unexplored approach to the aetiology and progression of osteoarthritis.

5. Ligament excision

A popular model for osteoarthritis is the excision of the anterior cruciate ligament (ACL) in the dog knee. The resulting radiographic changes are similar to that seen in humans. However, there is evidence that chondrocyte replication and proliferation of matrix is part of the natural history of cartilage changes in ACL deficient dogs (Brandt and Adams, 1989; Braunstein, Brandt and Albrecht, 1990). This is site dependent and means that careful evaluation is required if this model is used to test the efficacy of drugs for the induction of cartilage healing.

6. Immobilisation

Immobilisation of dog knees revealed a number of interesting changes in work by Jurvelin *et al.* (1989). The articular surfaces progressively softened due to decrease in proteoglycan content allowing increased permeability, without, interestingly, any evidence of chondrocyte death as is usually reported. Yet the articular surfaces appeared normal macroscopically. Mobilization for 15 weeks after 11 weeks' immobilization showed that whereas the patellofemoral joint was biomechanically normal throughout, the femoral and tibial condyles became softened by immobilization. Also, whereas the femoral condyle returned to normal after mobilization, the tibial condyle did not, maintaining a persistently high creep rate. The consequences of immobilisation were therefore site dependent. Chondrocytes thus adapt to the unloaded condition by decreasing proteoglycan synthesis. It was suggested that the ability to restore non-functional cartilage to normality would depend on an intact collagen network.

Chemical

1. Proteolytic enzymes

Proteolytic enzymes cause a diffuse superficial breakdown of the surface of cartilage via disruption of the organic matrix. They can therefore provide a good model for the synoviocyte-initiated type of damage seen in some cases of primary osteoarthritis. Papain in large doses is one method that has been described (Bentley 1971).

2. Antigen induced

This technique is used to simulate the immune effects of rheumatoid arthritis. One method uses ovalbumin in Freund's adjuvant as the antigen. An animal, usually the rabbit, is sensitized by an intradermal injection of the antigen, and is reimmunized 2 weeks later, and then given an intra-articular injection to provoke an arthritis (Pettipher *et al.*, 1989). This will again result in diffuse superficial damage.

Therapeutic manoeuvres

Drugs

Non-steroidal anti-inflammatory drugs (NSAI)

NSAIs are extremely effective in the symptomatic relief of arthritic pain. This is achieved by inhibiting cyclo-oxygenase production which converts arachnidonic acid into prostaglandins. This therefore reduces both the pain and swelling of synovial inflammation. Unfortunately, it also suppresses the inhibitory effect of prostaglandin on interleukin-1 (see Figure 2.2) and so accelerates cartilage breakdown (Pettipher *et al.*, 1989; Vignon *et al.*, 1990). A number of different mechanisms of action have, in fact, been reported depending on the NSAI used. They can promote or inhibit cartilage breakdown via cytokine production, matrix synthesis or degradation, or metalloprotease synthesis or suppression, with direct effects on both synoviocytes and chondrocytes. The goal has been to find an NSAI that has 'chondroprotection' (Doherty, 1989; Huskisson, 1990). Currently there is no NSAI that is truely chondroprotective, and it should be remembered that they also have complex effects on both muscles and bone metabolism (Doherty, 1989).

Aspirin inhibits cathepsin production from chondrocyte lysosomes. A beneficial effect of aspirin on induced superficial lacerations of the femoral groove in rabbits has been shown (Simmons and Chrisman, 1965; Ginsberg *et al.*, 1968). However, no effect was noted with its use in patients with chondromalacia patellae (Bentley, Leslie and Fischer, 1981). This could be explained on the grounds that the initiating insult, such as repetitive trauma, was still present in these patients, whereas in the rabbit experiments the insult was a single event (Ohno *et al.*, 1988).

Hormones

Steroids

The attraction of hormones in the treatment of cartilage degradation is that they could be used to switch on and maintain matrix synthesis by the chondrocytes. A further theoretical point is that the ingress of hormones is easier when there has been disruption of the superficial tangential layer of the cartilage. The use of steroids is strongly contested in the literature. Chondroprotection by corticosteroids has been indicated in rabbit and guinea pig experiments (Colombo *et al.*, 1983; Butler *et al.*, 1983; Williams and Brandt, 1985), and in humans (Pelletier *et al.*, 1987). But others have argued against their use in humans because of direct deleterious effects, and possible side effects (Salter, Gross and Hall, 1967; Behrens, Shepard and Mitchell, 1975). Pelletier and Martel-Pelletier (1989) showed that low-dose steroids (prednisolone and triamcolone hexacetonide) intra-articularly reduced osteophyte formation and the cartilage lesions in an ACL-deficient dog model of osteoarthritis. They argued for the use of low-dose long-acting steroids intra-articularly, the low dose used overcoming the reported adverse reactions. Steroids are thought to act directly on the chondrocyte and to inhibit metalloprotease production (see Figure 2.2). Pelletier and Martel-Pelletier concluded that research in this area is worth pursuing with newer steroids.

Growth factor peptides

Chondrocytes are known to have specific cell membrane receptors to various somatomedins that can induce chondrocyte replication, or stimulate organic matrix production (Osborn, Trippel and Mankin, 1989). Four peptides growth factors seem to be involved: insulin, fibroblast growth factor (FGF), epidermal growth factor (EGF), and somatomedin-C/insulin-like growth factor I (Sm-C/IGF-I). The results in animal experiments in mice, rabbits and cows, using monolayer cultures and organ cultures are still in a state of flux, but it seems that FGF alone can stimulate DNA synthesis, whereas the other three have to work in combination. The latter are thought to be progression factors, whereas FGF is a competence factor. This means that FGF renders the cell competent to proliferate (at the Go phase of cell division) and the others then allow progression to the S phase. Much work has still to be done before clinical work can start, but this is an exciting approach, with the prospect of switching on chondrocytes and thus inducing true cartilage healing.

Cross-linking agents

Since articular surface degradation is the first stage in the development of osteoarthritis, and is also the site of enzymic damage in rheumatoid arthritis, one approach has been to try to stabilize the damaged surface by inducing cross-linking of superficial layer collagen fibres. The agents used have been glutaraldehyde, osmium tetraoxide, and dithiobis succinimidyl proprionate (DSP). In vitro treatment of mouse femoral heads with these substances protected them against the action of clostridial collagenase (Stanescu and Stanescu, 1988). The femoral heads however, had normal articular cartilage. The effect on an arthritic model has not been described. Clinical experience with osmium tetraoxide used for chemical synovectomy in young active patients with pigmented villonodular synovitis (i.e. the articular cartilage is normal) has shown rapid and total articular destruction (H Dejour, personal communication). Probably the mechanical properties of normal cartilage were severely altered by increasing cross-linkage of the surface collagen causing rapid destruction under cyclical load.

Electrical

In 1974 Baker, Becker and Spadaro reported enhanced healing of osteochondral defects in immature rabbits using direct currents via implanted electrodes. Recently Lippiello, Chakkalakal and Connolly (1990) used pulsed direct current on adult rabbits and demonstrated enhanced repair compared with controls. This work also required implanted electrodes. There is no report of this type of work in humans.

Surgical

Although complete replacement of a joint by a prosthesis has been the greatest advance in orthopaedics, in a sense it is an admission of failure to preserve or return articular cartilage to normality. Despite the use of prostheses in younger patients, it is in this group that cartilage repair would have the most benefit, by avoiding multiple revision operations. A number of surgical manoeuvres could be used to reduce the rate of destruction or to enhance repair.

Excision and replacement

Excision of areas of superficial damage, and drilling of subchondral bone allows extrinsic repair to occur. Unfortunately, as has been remarked earlier, the resulting tissue is of poor quality and large defects will progress to osteoarthritis (Mitchell and Shepard, 1976). To overcome this attempts have been made to replace the defects.

Carbon fibre prostheses

In 1978 Forster *et al.* showed the fibroblastic response to carbon fibre in rabbits was not due to a toxic effect. Collagen patches for osteochondral defects were tried with some success (Speer *et al.*, 1979). Recently, carbon fibre patches have been used to fill large osteochondral defects. Since the normal extrinsic repair mechanisms are usually incomplete, the patch is used as a support to the fibrocartilaginous matrix (Muckle and Minns, 1990). Their animal work showed encouraging results 3 months from implantation with a smooth complete fibrocartilaginous surface. In 46 patients followed up for an average of 3 years they report a 75% excellent or good result. However, the defect is not filled with hyaline cartilage, so the later results may show a deterioration. They now want to impregnate the carbon patch with cultured chondrocytes (see below), and thus hope to promote hyaline cartilage production.

Synovium-fat-periosteum autograft

In a study on mature sheep from New Zealand (Rothwell, 1990) the femoral groove of the knee joint was denuded and the subchondral bone broached. A graft of synovium with its underlying fat and periosteum was taken from the medial side of the femoral condyle and sutured into the defect. The knees were immobilized for 6 weeks. One of the sheep was lost to follow up when it disappeared from the farm. Although the results up to 1 year were encouraging, at 2 years the grafts failed. No hyaline metaplasia took place. Other autologous tissues have been transplanted in the past with a similar lack of success (Rubak, 1982).

Cartilage replacement

A number of methods have been tried to replace chondral defects with cartilage. Chondrocytes may be cultured (Aston and Bentley, 1986; Wakitani *et al.*, 1989; Grande *et al.*, 1989), or cartilage itself used (Aston and Bentley, 1986; Homminga *et al.*, 1989). Rabbits are the favourite experimental animal. Aston and Bentley (1986) showed a 40% success rate with cultured chondrocytes from immature rabbits embedded in Ham's buffered medium with 10% fetal calf serum when implanted into defects in mature rabbits. Grande *et al.* (1989) modified this experiment by autochondrocyte cultures. Chondrocytes were cultured from the cartilage harvested when creating the defects in their rabbits. The cultures were reinserted under a periosteal flap. All 10 rabbits showed evidence of repair, which was complete in two, by 6 weeks. The repair tissue had predominantly type II collagen. Wakitani *et al.* (1989) embedded chondrocytes cultured from immature rabbits in a collagen gel and reported an 80% success rate with this modification.

Chondral defects can be grafted using articular cartilage blocks. Allografts from immature rabbits had an 80% success rate when transplanted into rabbits (Aston and Bentley, 1986). The grafts were held by a press-fit technique. Homminga *et al.* (1989) used autologous costal perichondrium and anchored this using human fibrin glue. They had an 85% success rate with hyaline cartilage formed in their successful cases.

Kaplonyi *et al.* (1988) have used fibrin glue in patients with traumatic chondral flaps. However, they drilled the subchondral bone as part of their technique, so the results are not comparable. They have also used this technique for osteochondral flaps (see below).

Small-fragment osteochondral allografts taken from organ banks have been used for some time in specialist centres (McDermott *et al.*, 1985). Seventy-five per cent were successful at 4 years in patients with traumatic lesions, 42% in patients with osteoarthritis. The results were assessed clinically and radiologically. The grafts were contoured and press-fitted into the defects.

Using fibrin adhesive in traumatic osteochondral and chondral flaps to anchor the flap and therefore avoid the use of metal fixation Kaplonyi *et al.* (1988) report a 75% success rate in 12 patients assessed arthroscopically. Immobilization for 6 weeks was required and some patients required additional internal fixation.

Large osteochondral allografts have been shown to have similar results at between 2 and 10 years follow up (Meyers, Akeson and Convery, 1989). There was a 77% success rate, falling to 30% in those patients who had grafts for unicompartmental osteoarthritis. Histological examination showed the grafts were alive.

Recently, Billings *et al.* (1990) have tried a composite osteochondral graft in rabbits. Demineralized bone matrix (DBM) allografts were anchored with vicryl sutures to autologous costal perichondral grafts. These were then press-fitted into osteochondral defects made in the medial femoral condyle. A control group of just DBM did as well as the composite grafts with an 85% 'repair' rate, but neither group produced hyaline cartilage at 12 weeks.

Cartilage grafting techniques have two main complications: infection (5–10%) and loss of graft (5%). Similar results are found in the clinical studies.

Mechanical realignment

The effect of mechanical realignment on arthritic joint articular cartilage has not been reported recently. Repair assessed arthroscopically has been reported after valgus upper tibial osteotomy for medial compartment osteoarthritis (Fujisawa, Masuhara and Shiomi, 1979). This type of repair was almost certainly extrinsic. With gross lesions intrinsic repair is unlikely. Reduction or halting the rate of degeneration is generally agreed to occur, provided the operation has been carried out expertly (Insall, Shoji and Mayer, 1974; Coventry, 1979).

Conclusion

Active research in the repair of damaged articular cartilage is currently going on in a number of areas. The old view that osteoarthritis is merely a mechanical problem of wear and tear has been replaced, in some quarters, with the view that the problem is entirely at a molecular level. This has led to the belief that osteoarthritis will ultimately be treatable by drugs alone. Reality is likely to lie somewhere between the two. With the success of total joint replacement, the field of induction of cartilage repair mechanisms and cartilage replacement is currently aimed at the younger patient with cartilage damage. A scheme for managing these patients is shown in Figure 2.3. In very

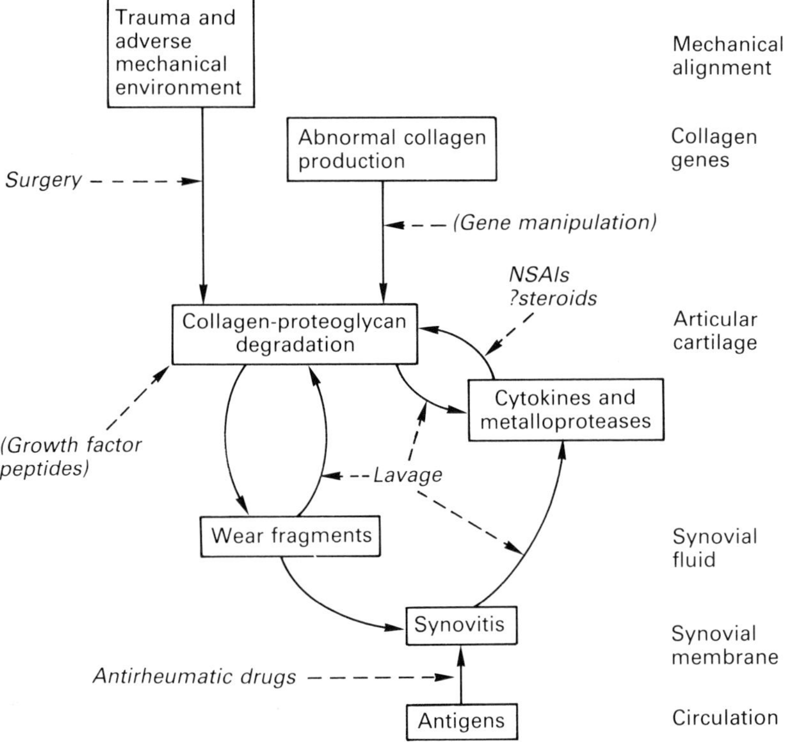

Figure 2.3 Early management of articular cartilage damage.

early cases, such as traumatic osteochondral defects, then a perfect surgical repair has a good chance of a perfect result. Likewise in rheumatoid arthritis, treatment initially can be entirely medical. However, the bulk of patients fall between the two treatment modalities. The way forward should be a combined effort with perhaps more use of drugs in the surgically-managed patients, and the judicious use of surgery, particularly lavage procedures, for those patients with primary idiopathic osteoarthritis. Since this condition may have a genetic element, the chances of a 'cure' are remote, and most patients will ultimately come to a joint replacement. Current research into slowing down the rate of cartilage degradation with drugs is therefore logical. However, experiments of osteoarthritis that use mechanical instability as models have doubtful value in the assessment of drugs except in demonstrating reduction in the rate of inevitable damage.

Once chondral damage exposes the subchondral bone then the chances of drug-healing are remote. The area of chondral replacement will, perhaps, yield a procedure that will help the younger patient. However, even in this group, a combined approach with chondral grafts and induction agents would seem to be the best way forward.

References

Aston, J.E. and Bentley, G. (1986) Repair of articular surfaces by allografts of articular and growth-plate cartilage. *Journal of Bone and Joint Surgery*, **68B**: 29–35.

Baker, B., Becker, R.O. and Spadaro, J. (1974) A study of electrochemical enhancement of articular cartilage repair. *Clinical Orthopedics*, **102**: 251–267.

Bentley, G. (1971) Papain-induced degenerative arthritis of the hip in rabbits. *Journal of Bone and Joint Surgery*, **53B**: 324–337.

Bentley, G. (1985) Articular cartilage changes in chondromalacia patellas. *Journal of Bone and Joint Surgery*, **67B**: 769–774.

Bentley, G., Leslie, I.J. and Fischer, D. (1981) Effect of aspirin treatment on chondromalacia patellae. *Annals of the Rheumatic Diseases*, **40**: 37–41.

Behrens, F., Shepard, N. and Mitchell, N. (1975) Alterations of rabbit articular cartilage by intra-articular injections of glucocorticoids. *Journal of Bone and Joint Surgery*, **57A**: 70–76.

Billings, E., von Schroeder, H.P., Mai, M.T. *et al.* (1990) Cartilage resurfacing of the rabbit knee. The use of allogeneic demineralized bone matrix-autogeneic perichondrium composite implant. *Acta Orthopedica Scandinavica*, **61**, 201–206.

Blake, D.R., Unsworth, J., Outhwaite, J.M. *et al.* (1989) Hypoxic reperfusion injury in the human inflamed joint. *Lancet*, **i**, 289–293.

Block, J.A., Inerot, S.E., Gitelis, S. and Kimura, J.H. (1991) Synthesis of chondrocytic keratan sulphate-containing proteoglycans by human chondrosarcoma cells in long-term cell culture. *Journal of Bone and Joint Surgery*, **73A**, 647–658.

Brandt, K.D. and Adams, M.E. (1989) Exuberant repair of articular cartilage damage: Effect of anterior cruciate ligament transection in the dog. *Trans. Orthop. Res. Soc.* **14,** 584.

Braunstein, E.M., Brandt, K.D. and Albrecht, M. (1990) MRI demonstration of hypertrophic articular cartilage repair in osteoarthritis. *Skeletal Radiology*, **19**, 335–339.

Bunning, R.A.D. and Russell, R.G.G. (1989) The effect of tumor necrosis factor α and γ-interferon on the resorption of human articular cartilage and on the production of prostaglandin E and of caseinase activity by human articular chondrocytes. *Arthritis and Rheumatism,* **32**, 780–783.

Butler, M., Colombo, C., Hickman L. *et al.* (1983) A new model of osteoarthritis in rabbits. III. Evaluation of anti-osteoarthritic effects of selected drugs administered intraarticularly. *Arthritis and Rheumatism*, **26**, 1380–1386.

Colombo, C., Butler, M., Hickman, L. *et al.* (1983) A new model of osteoarthritis in rabbits. II. Evaluation of selected anti-rheumatic drugs administered systemically. *Arthritis and Rheumatism,* **26,** 1132–1139.

Convery, F.R., Akeson, W.H. and Keown, G.H. (1972) The repair of osteochondral defects. An experimental study in horses. *Clinical Orthopedics,* **82**, 253–262.

Coventry, M.B. (1979) Upper tibial osteotomy for gonarthrosis. The evaluation of the operation in the last eighteen years and long-term results. *Orthopedic Clinics of North America*, **10**, 191–210.

Doherty. M. (1989) 'Chondroprotection' by non-steroidal anti-inflammatory drugs. *Annals of the Rheumatic Diseases,* **48**, 619–621.

DePalma, A.F., McKeever, C.D. and Subin, D.K. (1966) Process of repair of articular cartilage demonstrated by histology and autoradiography with tritiated thymidine. *Clinical Orthopedics,* **48**, 229–242.

Farkas, T., Lippiello, L., Mitrovic, D. *et al.* (1977) Papain induced healing of superficial lacerations in articular cartilage of adult rabbits. *Trans. Orthop. Res. Soc.,* **2,** 204.

Forster, I.W., Ralis, Z.A., McKibbin, B. and Jenkins, D.H.R. (1978) Biological reaction to carbon fiber implants. *Clinical Orthopedics,* **131**, 299–307.

Fujisawa, Y., Masuhara, K. and Shiomi, S. (1979) The effect of high tibial osteotomy on osteoarthritis of the knee. An arthroscopic study of fifty four knee joints. *Orthopedic Clinics of North America,* **10**, 585–608.

Ghadially, F.N., Thomas, I., Oryschak, A.F. and Laronde, J.M. (1977) Long term results of superficial defects in articular cartilage. *J. Pathol.,* **121,** 213–217.

Ginsberg, J.M., Eyring, E.J., Lacy, S. and Tomblin, W. (1968) Inhibition of cartilage destruction by intermittent salicylate. *Arthritis and Rheumatism*, **11**, 824–829.

Grande, D.A., Pitman, M.I., Peterson, L., Menche, D. and Klein, M. (1989) The repair of experimentally produced defects in rabbit articular cartilage by autologous chondrocyte transplantation. *Journal of Orthopaedic Research*, **7**, 208–218.

Homminga, G.N., van der Linden, T.J., Terwindt-Rouwenhorst, E.A.W. and Drukker, J. (1989) Repair of articular defects by perichondrial grafts. Experiments in rabbits. *Acta Orthopaedica Scandinavica*, **60**, 326–329.

Hull, R. and Pope, F.M. (1982) Osteoarthritis and cartilage collagen genes. *Lancet,* **i**, 1337–1338.

Huskisson, E.C. (1990) Clinical aspects of chondroprotection. *Seminars in Arthritis and Rheumatism,* **19** (Suppl. 1), 30–32.

Insall, J.N., Shoji, S. and Mayer, V. (1974) High tibial osteotomy. A five-year evaluation. *Journal of Bone and Joint Surgery*, **56A**: 1397–1405.

Jurvelin, J., Kiviranta, I., Saamanen, A.-M., Tammi, M. and Helminen, H.J. (1989) Partial restoration of immobilization-induced softening of canine articular cartilage after remobilization of the knee (stifle) joint. *Journal of Orthopaedic Research,* **7**, 352–358.

Kaplonyi, G., Zimmerman, I., Frenyo, A.D., Farkas, T. and Nemes, G. (1988) The use of fibrin adhesive in the repair of chondral and osteochondral injuries. *Injury*, **19**, 267–272.

Lippiello, L., Chakkalakal, D. and Connolly, J.F. (1990) Pulsing direct current-induced repair of articular cartilage in rabbit osteochondral defects. *Journal of Orthopaedic Research*, **8**, 266–275.

Mankin, H.J. (1982) Current concepts review. The response of articular cartilage to mechanical injury. *Journal of Bone and Joint Surgery,* **64A**, 460–466.

Mayne, R. (1989) Cartilage collagens. What is their function, and are they involved in articular disease? *Arthritis and Rheumatism,* **32**, 241–246.

McDermott, A.G.P, Langer, F., Pritzker, K.P.H. and Gross, A.E. (1985) Fresh small-fragment osteochondral allografts. *Clinical Orthopedics,* **197**, 96–102.

Meachim, G. (1963) The effect of scarification of articular cartilage in the rabbit. *Journal of Bone and Joint Surgery,* **45B**, 150–161.

Meyers, M.H., Akeson, W. and Convery, F.R. (1989) Resurfacing the knee with fresh osteochondral allograft. *Journal of Bone and Joint Surgery,* **71A**, 704–713.

Mitchell, N. and Shepard, N. (1976) The resurfacing of adult rabbit articular cartilage by multiple perforations through the subchondral bone. *Journal of Bone and Joint Surgery,* **58A**, 230–233.

Mow, V.C., Proctor, C.S., Kelly, M.A. (1989) Biomechanics of articular cartilage. In (Frankel, V.H. and Nordin, M. Eds.) *Basic biomechanics of the Skeletal System*, 2nd Ed. Lea & Febiger, Philadelphia; Ch 2; pp. 31–57.

Muckle, D.S. and Minns, R.J. (1990) Biological response to woven carbon fibre pads in the knee. A clinical and experimental study. *Journal of Bone and Joint Surgery,* **72B**, 60–62.

Nelson, B.H., Anderson, D.D., Brand, R.A. and Brown, T.D. (1988) Effect of osteochondral defects on articular cartilage. Contact pressures studied on dogs. *Acta Orthopaedica Scandinavica*, **59**, 574–579.

O'Hara, B.P., Urban, J.P.G. and Maroudas, A. (1990) Influence of cyclical loading on the nutrition of articular cartilage. *Annals of the Rheumatic Diseases,* **49**, 536–539.

Ohno, O., Naito, J., Iguchi, T. *et al.* (1988) An electromicroscopic study of early pathology in chondromalacia of the patella. *Journal of Bone and Joint Surgery*, **70A,** 883–899.

Osborn, K.D., Trippel, S.B. and Mankin, H.J. (1989) Growth factor stimulation of adult articular cartilage. *Journal of Orthopaedic Research*, **7**, 35–42.

Pelletier, J.P., Martel-Pelletier, M., Cloutier, J.M. and Woessner, J.F. (1987) Proteoglycan-degrading acid metalloprotease activity in human osteoarthritic cartilage, and the effect of intraarticular steroid injections. *Arthritis and Rheumatism*, **30**, 541–548.

Pelletier, J.P. and Martel-Pelletier, J. (1989) Protective effects of corticosteroids on cartilage lesions and osteophyte formation in the Pond-Nuki dog model of osteoarthritis. *Arthritis and Rheumatism*, **32**, 181–193.

Pettipher, E.R., Henderson, B., Edwards, J.C.W. and Higgs, G.A. (1989) Effect of indomethacin on swelling, lymphocyte influx, and cartilage proteoglycan depletion in experimental arthritis. *Annals of the Rheumatic Diseases*, **48,** 623–627.

Radin, E.L., Ehrlich, M.G., Chernack, R. *et al.* (1978) Effect of repetitive impulse loading on the knee joints of rabbits. *Clin. Orthop.*, **131,** 288–293.

Radin, E.L., Paul, I.L. and Lowy, M.A. (1970) A comparison of the dynamic force transmitting properties of bone and articular cartilage. *J. Bone Joint Surg.*, **52A,** 444–456.

Rothwell, A.G. (1990) Synovium transplantation onto the cartilage denuded patellar groove of the sheep knee joint. *Orthopedics*, **13**, 433–442.

Rubak, J.M. (1982) Reconstruction of articular cartilage defects with free periosteal grafts. *Acta Orthopaedica Scandinavica*, **53**, 175–180.

Salter, R.B., Gross, A. and Hall, J.H. (1964) Hydrocortisone arthropathy: an experimental investigation. *Canadian Medical Association Journal*, **97**, 374–377.

Shinmei, M., Okada, Y., Masuda, K., *et al.* (1990) The mechanism of cartilage degradation on osteoarthritic joints. *Seminars in Arthritis and Rheumatism*, **19** (Suppl. 1), 16–20.

Simmons, D.R. and Chrisman, O.D. (1965) Salicylate inhibition of cartilage degeneration. *Arthritis and Rheumatism*, **8**, 960–969.

Speer, D.P., Chvapic, M., Volz, R.G. and Holmes, M.D. (1979) Enhancement of healing in osteochondral defects by collagen sponge implants. *Clinical Orthopaedics*, **144**, 326–335.

Stanescu, R. and Stanescu, V. (1988) *In vitro* protection of the articular surface by cross-linking agents. *Journal of Rheumatology*, **15**, 1677–1682.

Vignon, E., Mathieu, P., Broquet, P., Louisot P. and Richard, M. (1990) Cartilage degradative enzymes in human osteoarthritis: effect of a non-steroidal antiinflammatory drug administered orally. *Seminars in Arthritis and Rheumatism*, **19** (Suppl. 1), 26–29.

Wakitani, S., Kimura, T., Hirooka, A. *et al.* (1989) Repair of rabbit articular surfaces with allograft chondrocytes embedded in collagen gel. *J. Bone Joint Surg.*, **71,** 74–80.

Watson, P., Kernohan, W.G. and Mollan, R.A.B. (1989) The effect of ultrasonically induced cavitation on articular cartilage. *Clinical Orthopaedics*, **245**, 288–296.

Williams, J.M. and Brandt, K.D. (1985) Triamcinolone hexacetonide protecs against fibrillation and osteophyte formation following chemically induced articular cartilage damage. *Arthritis and Rheumatism*, **28,** 1267–1274.

Chapter 3

The biology of bone

Phil L. Housden

Introduction

Bone is an important connective tissue in man and essential for many functions. Its macroscopic form may be divided into four basic categories, namely tubular bones, flat bones, short bones and other bones such as the vertebrae which do not fit into these categories.

Bone in all its various forms can be deemed to have three main physiological functions and the various bones with their different macroscopic and microscopic forms perform these functions to varying degrees. The roles are support, protection and bone mineral metabolism. The tubular bones are good examples of bones that aid in mechanical strength, with their facility to act as lever arms pivoting around the joints. The role of protection is obvious in the case of the flat bones of the skull but less so with the delicate marrow deep within the shafts of the long bones and within the sternum. The bone here allows the development of haemopoietic tissue in an environment which avoids squeezing or distorting the cells. Bone mineral metabolism tends to occur in the more vascular areas of bone such as the vertebral bodies and medullary cavities of some of the long bones where the bony structure is of a cancellous nature.

Bone metabolism is regulated by a number of humeral agents both local factors and hormones, details will be described. A description of the cells' origins and mode of function will be given with the constituent similarities and differences from cartilage being described in the chapter on cartilage.

History

The understanding of the structure and function of bone goes back many years but, throughout this time, increasing understanding of its nature has been brought about by improvements in techniques of observation and more recently labelling of bony tissue. Early microscopy of bone was performed by the Dutch amateur Anthony Van Leeuwenhoek (1632–1723). Magnification obtained at this time was between 40 and 200 times but some of the detail that could potentially have been described with early optical instruments was not clearly seen because of both poor illumination and poor specimen preparation. Van Leeuwenhoek did however describe the solid matrix of bone and also its penetration by a number of different 'tubuli'. These were probably haversian canals or possibly osteocytes. Clopton Havers in 1692 reiterated some of Van Leeuwenhoek's work and formally described the canals named after him (haversian canals). Purkinje in 1845 described the bone lacunae which he called bone corpuscles and Tomes and Morgan in 1853 postulated the idea of bone turnover introducing the concept of bone resorption. Queket in 1846 differentiated between calcified and uncalcified material and he appears to be the first person to have attempted to decalcify bone with the use of muriatic acid. Goodsir, at the same time described nucleated cells within the lacunae and these were later named osteocytes.

Osteoclasts were first described and named by Kolliker in 1873, and again Toms and Morgan demonstrated the function of 'minutely granular nucleated cells' in resorbing the endosteal surface of a long bone whose periosteal surface was exposed to the air in a dog model. They believed that as the bone lost substance the multinucleated cells increased, and this has subsequently been confirmed by Jowsey (1964).

Many technical advances have been made throughout this century both in the preparation of histological sections, their decalcification and also staining. The use of polarized light for inspecting bony sections has also proved to be useful since its introduction towards the end of the last century. Its unique properties are well suited for inspecting lamellae of bone as they contain parallel collagen fibres whose diffraction may be used to clarify imaging (Fig 3.1).

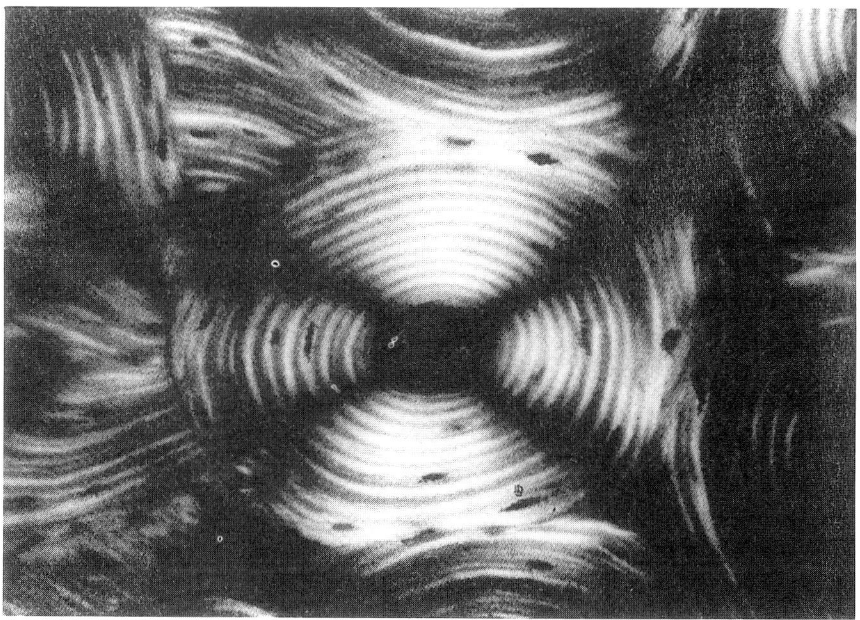

Figure 3.1 Lamella bone photographed under polarized light. (Reproduced from Gray's Anatomy 35th edn by kind permission of Churchill Livingstone, London.)

The techniques for preparing bone samples for analysis have improved in many areas. Preparation of fresh bone has always been difficult but, in recent years, the use of contact preparations has become useful. Sectioning of fresh bone is feasible, with care being taken to avoid mineral leaching in preparation, and with the final sample being set in a hard embedding medium, colloidin. Decalcification has become a more refined art although the resultant bone is still quite tough to cut. Dry bone can be ground down and, although this can give a good idea of the mineral architecture, the soft tissues and cellular architecture are obviously lost. Bone labelling with tetracycline (Frost 1969) has been used with its ability to fluoresce under ultraviolet light. Technitium labelling has been used in more recent years, with its affinity for osteoblastic cells and its ability to emit detectable gamma radiation. The use of microradiography in conjunction with microscopy and staining has also been found to be helpful and the mass spectrometer has been important in analysing the atomic make-up of bone mineral. The development of tissue cultures, which are still in their infancy, have further helped with the understanding of bone morphology and function. Immunospecific stains such as for alkaline phosphatase have been used to detect and monitor the activities of different cell types. In clinical practice labelling for alkaline phosphatase is useful to determine the origin of anaplastic cells and may help to differentiate between anaplastic osteosarcoma and a high grade chondrosarcoma (the latter being unstained).

The evolution of bone

Substances similar to lamellean bone were believed to have existed some 400 million years ago in early vertebrates. An armoured fish of the genus ostracoderms was believed to have had a plate-like structure surrounding it. This is believed to have had three layers, an inner one with parallel lamellae, a middle one of spongy type 'bone' and an outer one with tooth-like elevations on it formed from a substance not dissimilar to tooth enamel. A close relative of this fish, *Ostracodermis heteracia*, is believed to have had a similar, although acellular, shell.

On the face of it this may appear to suggest that bone has developed from this acellular 'aspidin' but it should be noted as Kolliker (1859) pointed out that some of the higher species of teleosts (fish) have developed acellular bone apparently from cellular ancestors, and it has been questioned which is a more advanced form of skeleton.

The origin of collagen is not clearly understood, but it is possibly developed from a unicellular organism where it may have been a surface protein that regulated metabolism and subsequently diversified its role into one of support. It certainly evolved early as a similar form is found in porifera, coelentera and insects which represent a broad spectrum of distantly related phyla. Little is known of the evolution of the ground substance of bone.

Development

The skeleton in the fetus is derived from mesenchymal cells in the sandwich between endodermal and ectodermal cells. The axial skeleton forms from the fragment of the somites known as the scleroderm. The appendicular skeleton is formed from outpouchings of the embryo as the primitive limb buds with the mesenchymal cells differentiating to precartilage and laying down the cartilage model which is subsequently surrounded by vascularized perichondrium and forms the primitive cortex. Bands of cartilage form at each end of the primitive long bones as epiphyseal plates and the perichondrial vascularised tissue permeates the plates. After a variable time an ossification centre appears in the epiphyseal cartilage. The development of longitudinal and circumferential growth is then allowed by the growth plate (epiphyseal plate) which is shown below (see Fig. 3.3). The metaphysis and epiphysis become fused once the epiphyseal plate ossifies and this is under the regulation of systemic hormones whose details are not fully understood.

Bone and its structure

The macroscopic structure of bone is determined primarily by inheritance with subsequent morphology being influenced by function and environmental influences.

The microstructure of bone can be divided into:

1 non-lamella bone, that is early woven or coarse fibred bone which has coarse collagen bundles within it and is the form of bone in its embryonic state, in callus or in isolated areas of adult bone.
2 lamella bone, this constitutes most mature bone and it is found to contain fine parallel fibres of collagen and is the form of bone found in adult cortical and cancellous bone.

During development bone forms either:

1 intramembranously, with direct transformation of mesenchymal tissue or
2 intracartilaginous (endochondral), where a cartilage model of the bone is replaced by initially coarse fibred and then lamella bone.

Structural elements

Bone is a complex organ but, as with cartilage, its essential structure can be divided into four components:

1 A protein scaffold which is formed from type 1 collagen.
2 An amorphous ground substance which is formed of other proteins with carbohydrate moities attached (glycoproteins).
3 A crystalline structure, the majority of which is calcium hydroxyapetite which is insoluble at the physiological pH of 7.4.
4 Bone cells, namely osteoblasts, osteocytes and osteoclasts.

Mature adult bone is 20% water, it contains 30 or 40% organic tissue (90% collagen) and when examined in the dry state 60–70% of its weight is made up of mineral. Looking in more detail at the orientation of these different constituents bone can be divided into lamella and non lamella or woven bone.

Woven bone

This has been described as being like a gruyère cheese consisting of bony trabeculae with relatively coarse collagen fibres coursing through them in all directions in a manner similar to carpet underlay. Due to the random pattern of fibres, its imaging is not improved by polarized light. Osteocytes are randomly distributed within the bone matrix and have cytoplasmic projections passing down canaliculi. The ground substance is a polysaccharide protein matrix that is believed to be more metabolically active than in lamella bone and tends to have a more strongly positive periodic acid shift (PAS) reaction implying less polymerization of the proteins. The woven bone tends to be the earlier bone to be laid down, it is found in embryonic tissue and also occurs at sites of healing fractures in the form of callus. This type of tissue is also found in osteosarcomatous lesions and in areas of ectopic ossification. In humans, the long bones, which have significant quantities of woven bone in them at birth, replace it with lamella bone by the age of 2 or 3 years.

Lamella bone

This is a more organized tissue with finer parallel collagen fibres coursing through it. The ground substance again is of protein and polysaccharide with the mineral crystals permeating between the fibres. As the name suggests the tissue is laid down in layers which are some 5–7μm thick and these are formed in one of three arrangements, (Figure 3.2).

1 Small rings not dissimilar to a bisected onion; this is the form of the haversian system with a vascular canal at the centre which is some 50μm across. These are also known as osteons.
2 Interstitial lamellae are seen as stacked plates and are believed to be the remnants of previous haversian systems, the majority of which has been resorbed.
3 Circumferential lamellae running just below the periosteum and wrapping around the circumference of a long bone.

In all cases each successive lamella has collagen fibres running within it, parallel to one another but orientated approximately 90° to the layers on either side. Because of this arrangement polarized light can be used to observe the interface betwen the lamellae, (See Figure 3.1).

Lacunae are found within lamella bone appearing regularly throughout the haversian systems with interconnecting canaliculi running in a plane perpendicular to the axis of the vascular core. There has, however, been described no specific position of the lacunae with respect to the lamellae and cement lines.

The ground substance in this bone is believed to be less reactive and more polymerized with a less intense PAS reaction.

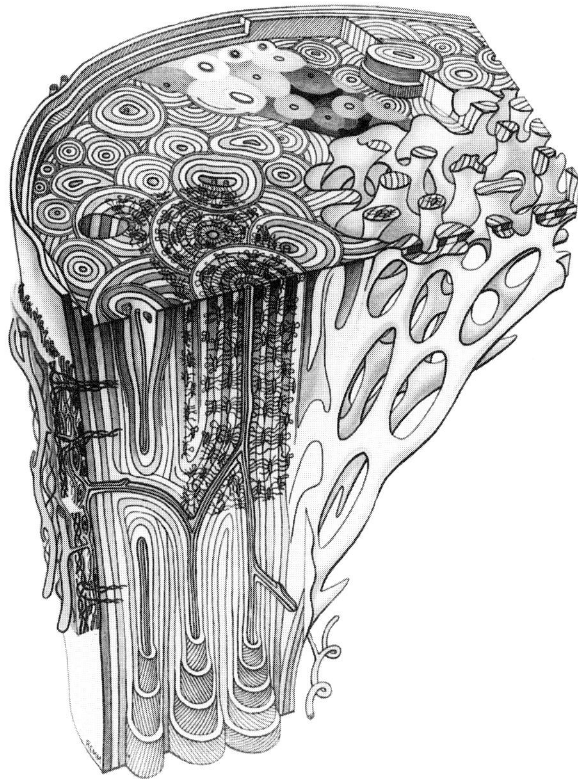

Figure 3.2 Gross structure of adult bone. (Reproduced from Gray's Anatomy 35th edn by kind permission of Churchill Livingstone, London.)

The epiphyseal growth plate

The microscopic appearance of the epiphyseal growth plate is shown in Figure 3.3 with four layers described. The reserve zone at the epiphyseal end of the plate has an unknown function, although cells from here may migrate into the proliferating zone. Cells in the proliferating zone (labelled growth) are found in columns and the quantity of endoplasmic reticulum increases in their cytoplasm by about three times as they move away from the reserve zone. This is the only layer where cell division has been demonstrated. The third zone is the hypertrophic zone (labelled transformation) where the cells increase in size by about five times. Glycogen which is present in the cytoplasm in the cells of the proliferation zone is lost and cytoplasmic granules appear in the cells as they migrate towards the zone of provisional calcification (labelled ossification). These granules contain calcium and phosphate. In the upper part of the hypertrophic zone calcium is found in the mitochondria but this appears to migrate into the cytoplasmic granules by the time the cells reach the zone of provisional calcification. Calcification occurs within the zone of provisional calcification, initially in the interstitial spaces between the columns. The understanding of calcification of bone tissue has developed very rapidly in the last 10 years when the concept of the

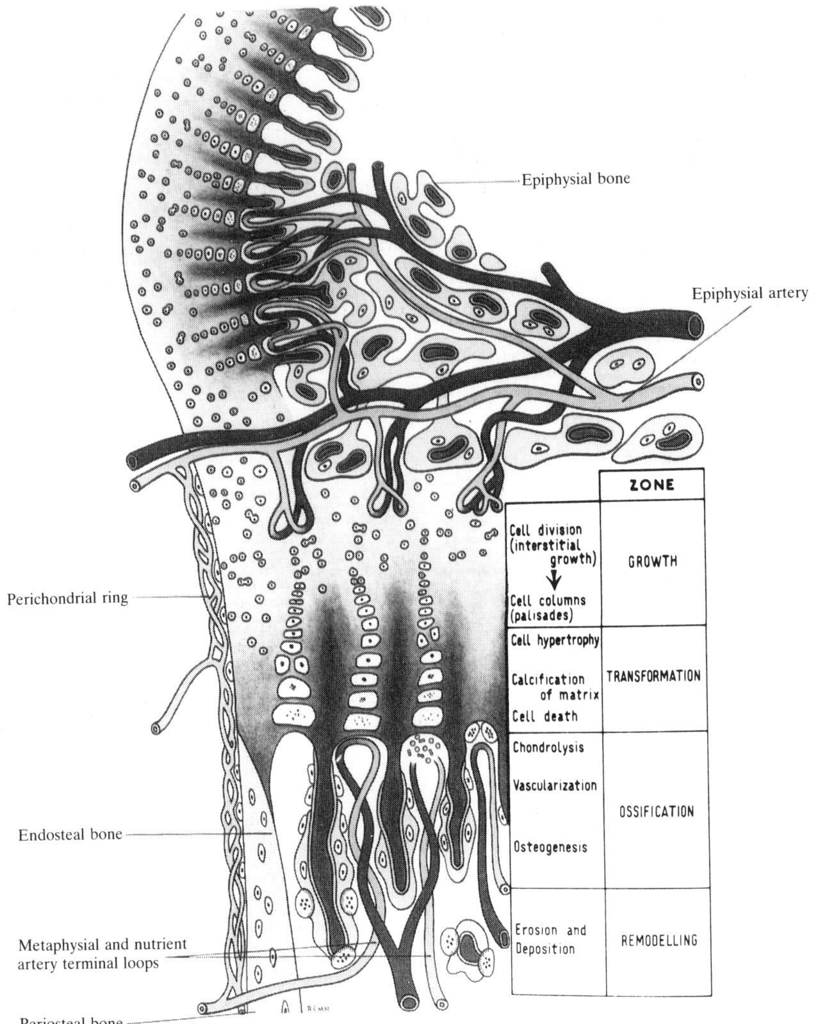

Figure 3.3 The epiphyseal growth plate. (Reproduced from Gray's Anatomy 35th edn by kind permission of Churchill Livingstone, London.)

matrix vesicle was first introduced (Anderson, 1976; 1978; 1985); this will be described in more detail later. Below the zone of provisional calcification is the metaphysis which is well vascularized. The upper part of this region contains active osteoblasts which initially produce woven bone within the calcified cartilage that is subsequently replaced by the definitive lamella bone. Osteoclasts are also present within the metaphysis and are involved in remodelling as growth proceeds.

A groove has been described indenting the margin of the interface between the reserve and the proliferation zone, the ossification groove, and this contains cells which are believed to pass into the margin of the epiphyseal plate to allow lateral growth to occur along with the longitudinal growth described above.

Bone cells, their origin and function

Previous authors believed that the origins of osteoblasts, osteocytes and osteoclasts were from similar cell lines. Recent work has refuted this and it is now believed that the osteoblasts develop from local stromal cells and occasionally from fibroblasts which are specifically induced (Fig. 3.4). The osteoclasts with their characteristics similar to multinucleate giant cells are believed to be derived from haemopoietic cells in a similar manner to the macrophages. Osteoblasts during the growth of bone become incorporated into the osseous tissue and once this occurs their cytoplasmic architecture appears to change in tandem with their function and they become osteocytes.

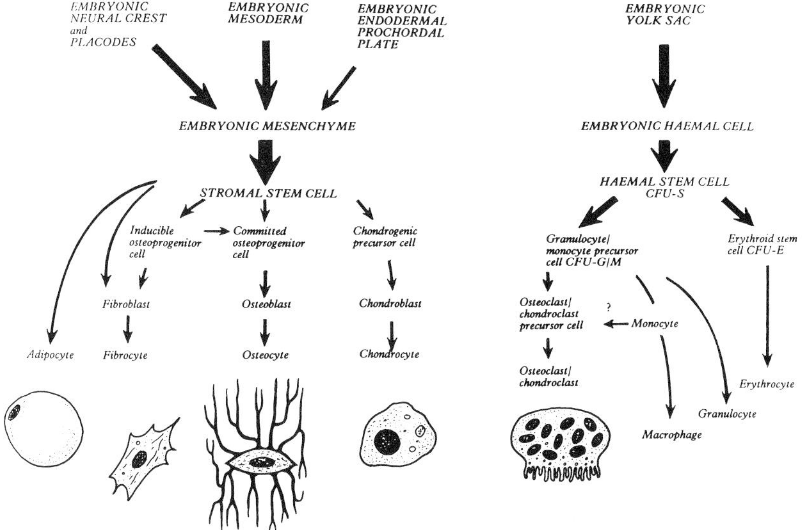

Figure 3.4 Bone cells, their structure and function. (Reproduced from Gray's Anatomy 35th edn by kind permission of Churchill Livingstone, London.)

Osteoblasts

These cells have characteristics similar to fibroblasts and indeed, have similar origins, but there is debate about what stimulates fibroblasts to develop osteogenic potential. The osteoblasts have been shown to contain large amounts of alkaline phosphatase and this differentiates them histologically from simple fibroblasts. The factors which change a fibroblast into an osteoblast are ill understood, but there are believed to be determined and inducible cells; the latter have the facility under suitable local and systemic conditions to differentiate into osteoblasts as is believed to be the case in the formation of callus around a fresh diaphyseal fracture. The induction depends particularly on local conditions and substances which have been suggested include bone morphogenic protein which is believed to stimulate the mesenchymal cells to become chondrocytes or osteoblasts, (Urist *et al.*, 1982).

Osteoblasts are found covering the surface of bone matrix and have nuclei which stain strongly for RNA with basophilic cytoplasm, they also stain strongly for alkaline phosphatase. When forming bone the osteoblasts produce collagen (type 1) which is formed in the rough endoplasmic reticulum and deposited via the golgi apparatus extracellularly. These cells also produce the protein and polysaccharide ground substance. The uncalcified bone matrix thus produced is known as osteoid. Calcification of osteoid is initiated by the matrix vesicles as described below and so bone is formed.

Calcification of osteoid

The process of calcification has been elucidated in some detail in recent years. Calcium and phosphate deposition has been shown initially to occur in extracellular vesicles known as matrix vesicles which are membrane bound extracellular bodies 100–200 nm in diameter (Fig. 3.5). These membrane bodies contain alkaline phosphatase, a large quantity of lipid to which calcium tends to bind, and have an intact membrane which may have a calcium pump within it (Anderson, 1976; 1978; 1985).

Once crystallization has been initiated, matrix vesicles are believed to be degraded by both enzymic and mechanical factors. The origin of the membrane vesicles has not been precisely determined but it appears most likely that they bud off from the plasma membrane of the osteoblasts. The propagation of mineralization is obviously dependent on relatively high levels of calcium and phosphate ions and it appears that the calcium, which is concentrated in mitochondria of the osteoblasts in the proliferating zone of the growth plate, may subsequently be released and precipitate around the crystals of the matrix vesicles.

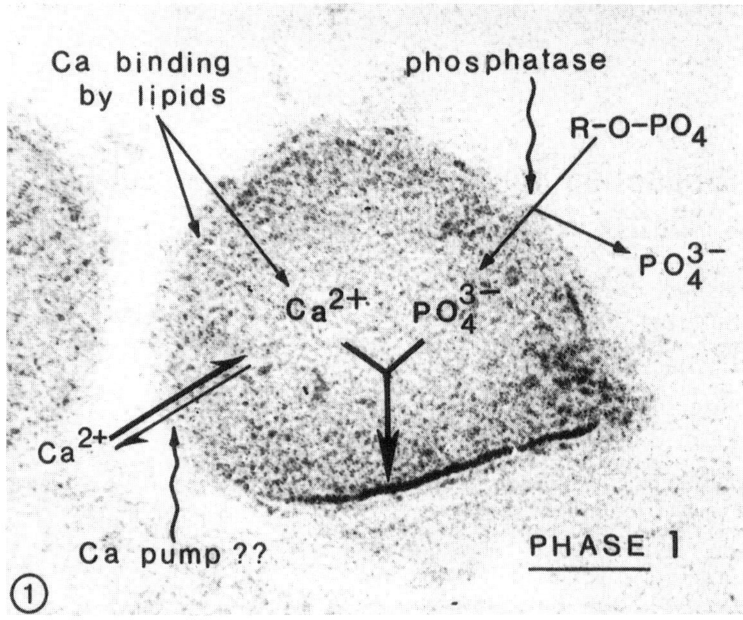

Figure 3.5 A matrix vesicle under electron microscopy. (Reproduced with permission from Anderson, H.C.)

Osteoblasts which are inactive have less basophilic cytoplasm and tend to be flatter. They cover the entire surface of the bone matrix, both periosteally and endosteally, the only exception being areas where osteoclasts are active. Interestingly proteases have been identified in the cytoplasm of osteoblasts capable of matrix degradation. Their role is not clear but, significantly, there is no evidence to date that osteoblasts are involved in bone demineralization.

Osteocytes

It has been estimated that about 10% of osteoblasts become osteocytes as they are incorporated into the osteoid and mineral structure. They lie in lacunae, and are connected by fine canaliculae to each other and the surface osteoblasts. To date the exact role of osteocytes is not clear.

It appears they can reabsorb and replace bone and may do this as part of the removal and replacement of perilacular bone possibly being linked to calcium homeostasis. Variable sizes of the lacunae around osteocytes have been observed depending on the prevailing levels of calcium and phosphate. This observation supports the view that osteocytes may be important in moderating short-term fluctuations in serum calcium and phosphate levels.

Osteoclasts

The function of osteoclasts is principally, and possibly solely, the resorption of bone. This is carried out with the osteoclasts closely adherent to the bone matrix and two interface zones are described. The first is known as the clear zone and this contains actin-like filaments and the cytoplasm has otherwise little else in it. The clear zone is believed to facilitate adherence of the osteoclasts to the surface of the bone matrix. The second zone known as the A zone (Hancox and Boothroid, 1972) is characterized by a ruffled border with infoldings of the cytoplasm and is believed to be the active resorbing site. Free calcium hydroxyapatite crystals have been seen in cytoplasmic vacuoles suggesting a phagocytotic means of crystal digestion. No collagen remnants have been identified within the cytoplasm of the osteoclasts and their breakdown is believed to be purely extracellular. The extracellular space, which includes the infolding channels, has been shown to contain significant levels of hydrolytic enzymes including acid phosphatase. The concentration of this is increased by parathyroid hormone and decreased by calcitonin. The collagen fibrils which have been found in the extracellular space in the A zone are deeply argyrophilic, a characteristic similar to fine relatively soluble reticulin fibres, in contrast the usual type 1 collagen which is common in bone is practically unstained by silver. This is most suggestive that the osteoclasts may excrete some form of collagenase which disrupts the cross-linking between bone matrix collagen fibres and thus alters its staining characteristics and similarly its solubility.

Regulation of Osteoblastic and Osteoclastic Activity

It is now generally accepted that the osteoblasts and osteoclasts perform their functions in close coordination with one another and the concept of a bone remodelling unit (BMU) has been described. The activity of the two cell types of the unit are believed to be regulated by a number of factors, most of which are ill understood. The regulation mechanisms may be either through local mediators or systemic hormones. Although the idea of coupling is now well accepted, the local factors responsible for changes in activity are not known.

Systemic hormones

Parathyroid and vitamin D are the two well known systemic hormones involved in calcium metabolism along with calcitonin, whose physiological role has been questioned in recent years. Calcitonin, however, probably has a regulating role and this is supported by the suggestion by Taggart *et al.* (1982) that calcitonin receptors on osteoclasts may be deficient in osteoporosis.

Parathyroid Hormone (PTH)

Parathyroid hormone is the most important regulator of calcium metabolism. If the systemic level of free ionized calcium drops, parathyroid hormone production is stimulated and this in turn increases the renal tubular resorption of calcium and excretion of phosphate, the production of active vitamin D (1,25 dihydroxycholecalciferol) or calcitrol and the mobilization of mineral from the bone by stimulating osteoclastic activity. This latter function is reflected in the increase in size of the ruffled border of these cells which appears to be proportional to their activity. However, the hormone action appears to be an indirect one dependent on the presence of osteoblasts, with no PTH receptors having been found on osteoclasts (Jilka, 1986). Parathyroid hormone also directly influences calcium and phosphate absorption from the intestine, although part of this activity is through its activation of calcitrol. The effect of parathyroid hormone on osteoblasts is variable and is dependent on the dose and duration of treatment. In vivo, parathyroid hormone has been shown directly to inhibit the production of osteoid collagen and thus decrease the volume of osteoid to be mineralized. This, in tandem with its effect on osteoclasts, would appear logically to fit into a haemostatic mechanism maintaininng a satisfactory level of extracellular calcium. It is interesting, however, that patients treated with parathyroid hormone who have hypoparathyroidism do in fact increase their trabecular bone volume implying that it has an anabolic effect on bone, but this is probably an indirect effect, possibly mediated by a second hormone.

Vitamin D

Some 15 natural metabolites of vitamin D have been isolated, but the physiologically active hormone appears to be the 1,25 dihydroxycholecalciferol. Active vitamin D (calcitrol) has an important direct role on increasing calcium and phosphate absorption from the small intestine, but it also acts in concert with parathyroid hormone to mobilize calcium from bone. Vitamin D appears to have a similar effect on bone to parathyroid hormone (both anabolic and indirectly catbolic)

Calcitonin

This hormone is known to have a direct inhibitory effect on bone resorption but does not affect osteoblasts directly. It is now believed by many not to be an important physiological regulator of bone metabolism in humans, although it does have a significant effect which can be used therapeutically in unphysiological doses. However, failure of inhibition of osteoclast receptors has been implicated in the development of osteoporosis (Dixon *et al.*, 1983; Taggart *et al.*, 1982).

Glucocorticoids, insulin and other hormones

The glucocorticoids have varied effects on bone metabolism. Their well known osteopenic effect is believed to be an indirect action and indeed experiments on osteoblasts in culture have suggested a direct stimulant effect on osteoblastic cells.

Insulin has been investigated for its effects on bone metabolism, but to date no important direct effect has been found. The incidence of osteopenia in poorly controlled diabetics would appear to be an indirect phenomenon. Growth hormone, somatomedin, thyroid hormone, the androgens and oestrogens have all been implicated in bone metabolism but their direct effects have been hard to determine due to the difficulty in devising cell models on which to test them. A summary of the effects of these hormones is given in Table 3.1.

Table 3.1 Probable effects of some hormones on bone cell function

	Effect on osteoblast activity	
	Direct	*Indirect*
Parathyroid	−	+
Vitamin D (1,25 dihydroxycholecalciferol	−	+
Calcitonin		+
Growth hormone		+
Glucocorticoids	?	−
Insulin	+	+
Thyroxine		+
Somatomedin	+	+
Platelet derived growth factor	+	?
Ions		
Calcium	+	+
Phosphate	+	+

Local factors

Much bone metabolism is regulated centrally by many of the hormones described above but, it has become evident that there must be local control, probably through humoral factors, which effect the bone remodelling units. This becomes obvious when considering potential regulating mechanisms for fractures or immobilized limbs where obvious limited local changes in osteoblastic and osteoclastic activity occurs. A number of local 'hormones' have been described including prostaglandins, osteoclastic activating factor, bone morphogenic protein and bone determinant protein. Prostaglandins, particularly of the E series are known to be potent stimulators of bone resorption, and have been implicated in osteolysis and hypercalcaemia of malignancy. Interestingly they do not appear directly to inhibit osteoblastic collagen production in doses which effect bone resorption, and in fact children treated with prostaglandin to maintain a patent ductus arteriosus have been shown to have marked increases in periosteal reaction. Prostaglandin E2 has been found in osteoblasts and may well be a local coupling agent.

Osteoclastic activating factor appears to be produced locally by lymphocytes and stimulates osteoclasts. This may be the factor that initiates lytic bone lesions in lymphoma and myeloma (Mundy *et al.*, 1974).

Bone morphogenic protein (BMP) has been described by Urist (1982) and he believes that this has a molecular weight of some 17 000–18 000. Around the same time, Drivdahl (1982) described bone derived growth factor (BDGF) and he believes this to be a different protein. BDGF is believed to be an important coupling factor with a stimulant effect on osteoblasts. It is believed to have molecular weight of over 12 000 and it remains to be seen whether these two substances are in fact the same. Further to this Raisz *et al.* (1983) have described two different bone derived growth factors, namely 1 and 2!

Calcium phosphate and other ions as regulators

Due to its constant concentration, calcium is unlikely to be a direct regulator of bone metabolism, although an adequate serum concentration is essential for bone formation. Phosphate, on the other hand, has a wide physiological range and it is possibly implicated as a regulator, although this is controversial. Acid–base balance is known to have marked effects on bone mineral metabolism as is demonstrated in patients with renal tubular acidosis in whom skeletal growth is markedly impaired. Finally, magnesium lack has been implicated in impaired growth of the skeleton, although the mechanism is ill understood, as are the effects of other ions on skeletal growth.

Electricity and mechanical effects

The distortion of many materials including vital structures such as bone and muscle has been shown to produce alterations in electrical potential (the piezo electric effect). The electromagnetic effect thus produced when a fracture occurs with bone distortion has been implicated as a stimulus or intermediate messenger in the stimulation of new bone formation. There is controversy as to whether electromagnetic fields around fractures can speed up the production of callus, although work from Sharrad (1990) certainly supports this hypothesis.

Conclusion

Much is still to be learnt about bone, its function and metabolism. Areas where advances are hoped to be made are in laboratory cell cultures, where specific direct effects of hormones and factors may be observed.

In recent years significant areas of improved understanding include the confirmation that osteoblasts and osteoclasts have separate stem cells. The discovery of matrix vesicles, possibly originating from osteoblasts, and their involvement in mineralization has been elucidated. The derivation of osteoclasts from the macrophage cell line is confirmed, and the intracellular degradation of bone mineral with extracellular degradation of collagen seems accepted.

Finally, bone hormone control mechanisms have been studied. The indirect effect of parathyroid hormone on osteoclasts via osteoblasts is interesting, confirming the intimate relationship of these cells in the bone modelling unit, although specific substances that act as local factors have, to date, not been identified. The stimulant effect of a magnetic field on fracture healing is probable and once a usable humoral intermediate of the osteoblast mechanism can be identified there may be further advances in the treatment and avoidance of delayed and non-unions of fractures.

References

Anderson, H.C. (1976) Matrix vesicle calcification. *Federation Proceedings,* **35**, 105–108.

Anderson, H.C. (1976) Introduction to the second conference on matrix vesicle calcification. *Metabolic Bone Disease and Related Research,* **1**, 83–87.

Anderson, H.C. (1985) Matrix vesicle calcification, review and update. *Bone and Mineral Research (3)* (ed. W.A. Peck), Ch. 6, pp. 109–139, Elsevier Science Publishers B.V.

Dixon, A.S. and J. (eds) (1983) Osteoporosis; a multidisciplinary problem. *International Congress and Symposium Series,* no. 55. Royal Society of Medicine, London; Academic Press; Grune and Stralton.

Drivdahl, R.H. (1982) Extracts of bone containing a potent regulator of bone formation. *Biochemica et Biophysica Acta,* **714**, 26–33.

Friendenstein, A.J., Petrakova, K.V., Kurolesova, A.I. and Frolova, G.P. (1968) Heterotopic transplantation of bone marrow. *Transplantation*, **6**, 230–247.

Frost, H.M. (1969) Tetracyclin–based histological analysis of bone remodelling. *Calcified Tissue Research,* **3**, 211–237.

Hancox, N.M. (1972) *The Biology of Bone.* Cambridge University Press.

Jilka, R.L. (1986) Are osteoblastic cells required for the control of osteoclastic activity by P.T.H. *Bone and Mineral,* **1**, 261–266.

Jowsey, J. (1964) Variation in bone mineral with age and disease. In *Bone Biodynamics* (ed. H.M. Frost) 461–479, Little Brown & Co.

Kölliker, A. (1873) Die normale resorption des knochengewebes und ihre Bedeutung für die Entstehung der typischen knochenformen, Leipzig, Volgel.

Mundy, G.R., Raisz, L.G., Cooper, R.A., Schechter, G.P. and Salmon, S.E. (1974) Evidence for the secretion of an osteoclast activating factor in myeloma. *New England Journal of Medicine,* **291**, 1041–1046.

Raisz, L.G., Kream, B.E. (1983) Regulation of bone formation (two parts). *New England Journal of Medicine,* **309**, 29–33; **309**, 83–89.

Sharrard, W.J. (1990) A double-blind trial of pulsed electromagnetic fields for delayed union or tibial fractures. *Journal of Bone and Joint Surgery,* **72B** (3) 347–55.

Taggart, H., Chestnut, C.H., Ivey, J.L., Baylink, D.J., Sisom, K. and Huber, M.B. (1982) Deficient calcitonin response to calcium stimulation in post menopausal osteoporosis. *Lancet*, i, 475–478.

Urist, M.R., Lietze, A., Mizutani, H., Takagi, K., Triffitt, J.T., Amstutz, J., de Lange, R., Termine, J. and Finerman, G.A. (1982) A bovine low molecular weight bone morphogenic protein fraction. *Clinical Orthopedics,* **162**, 219–231.

Further reading

Andrew, C. and Bassett, L. (1968) Biological significance of piezoelectricity. *Calcified Tissue Research,* **1**, 252–272.

Ashton, B.A. (1985) Characterisation of the cells with high alkaline phosphatase activity derived from human bone marrow. *Bone,* **6**, 313–319.

Baylink, D.J. and Wergedal, J.E. (1971) Bone formation by osteocytes. *American Journal of Physiology*. **221**, 669–678.

Bonucci, E. (1981) New knowledge on the origin function and fate of osteoclasts. *Clinical Orthopedics,* **158**, 252–271.

Brighton, C.T. (1978) Structure and function of the growth plate. *Clinical Orthopedics,* **136**, 22–31.

Chambers, T.J. (1980) The cellular basis of bone resorption. *Clinical Orthopedics,* **151**, 283–293.

Dodds, R.A. (1989) Comparative metabolic enzymic activity in trabecular as against cortical osteoblasts. *Bone,* **10**, 251–254.

Ganong, W.F. (1983) *Medical Physiology*, 11th edn. Appleton & Lange, Connecticut.

Grants Atlas of Anatomy, (1983) 8th edn. Williams and Wilkins, Baltimore.

Gray's Anatomy, (1973) 35th edn. Churchill Livingstone, London.

Hagenaars, C.E. van der Kraan, A.A.M., Kawilarang-de Haas, E.W.M., Visser, J.W.M. and Nijweide, P.J. (1989) Osteoclast formation from cloned pluripotent haemopoietic stem cells. *Bone and Mineral*, **6**, 179–189.

Owen, M. (1985) Lineage of osteogenic cells and their relationship to the stromal system. *Bone and Mineral Research,* (ed. W.A. Peck) **3**, Ch. 1, 1–17 Elsevier Science Publishers B.V.

Rasmussen, H. (1978) Vitamin D and bone. *Metabolic Bone Disease and Related Research,* **1**, 7–13.

Revell, P.A. (1985) *Pathology of Bone*. Springer-Verlag, New York.

Wong, G.L. (1986) Skeletal effects of P.T.H. *Bone and Mineral Research* (ed. W.A. Peck) (4), Ch. 3, pp. 103–119 Elsevier Science Publishers B.V.

Chapter 4

The biology of muscle and nerve injury

Allan N. Stirrat

Introduction

The subject of orthopaedics is generally conceived by the lay public and the medical profession as concerning the management of disorders of bone and joint. In the field of traumatic injury, care of the soft tissues, that is skin, muscle, blood vessels and nerves, is paramount. Muscle and nerve may be injured in isolation or in association with fractures, particularly those that are widely displaced and open. It is vital to remember that indirect injury to muscle and nerve may occur due to ischaemia and pressure. Unfortunately, a significant number of injuries to nerves are precipitated by surgeons (Birch *et al.*, 1991; Bonney 1983, 1986).

The aim of this chapter is to relate scientific knowledge to the processes of muscle and nerve repair and to review the prospects of improved outcome as a result of laboratory and clinical advances.

Skeletal muscle injury

Muscle is composed of cells known as fibres. Each fibre or cell has hundreds of nuclei. The fibres are arranged in fasciculi. Within each cell are situated the myofibrils and their associated mitochondria and sarcoplasmic reticulum. Each myofibril is composed of sequential scarcomeres containing the contractile elements of actin and myosin (Fig. 4.1). Skeletal muscle contraction is dependent on electrical events commencing in the central nervous system, disseminating through the peripheral nervous system and finally triggering neuromuscular transmission by release of the neurotransmitter, acetylcholine. Release of this transmitter produces a depolarizing end-plate potential which propagates through the muscle fibres causing muscle contraction.

The motor unit is the functional subdivision of muscle and comprises a single motor neurone and a group of muscle fibres varying from six to two thousand per motor unit. In general, the number of fibres is inversely proportional to the degree of fine control required of the muscle. For example, intrinsic muscles of the eye have few fibres in each motor unit. It should be noted that the fibres of the motor unit are not necessarily adjacent and may be scattered throughout the muscle. This provides for uniformity of

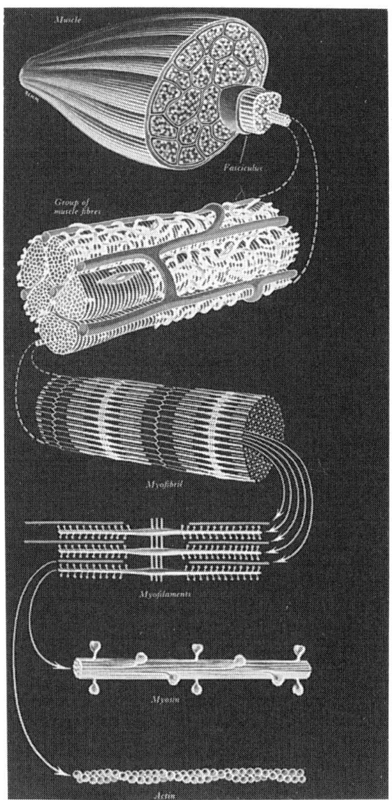

Figure 4.1 Diagram showing successive levels of organization within a skeletal muscle, from whole muscle, through fasciculi, fibres, myofibrils, myofilaments, down to molecular dimension. (Reproduced from Gray's Anatomy, 37th edn, 1989 by kind permission of Churchill Livingstone, London.)

contraction. Humans are provided with two types of muscle fibres, type I or slow fibres, and type II or fast fibres (Table 4.1).

Type I fibres tend to be situated in muscles that require steady prolonged contraction, for example, postural muscles of the spine. Type II fibres are located in those muscles involved in rapid powerful movements. Four types of fast fibres have been discovered, termed A, B, C and M (American Academy of Orthopaedic Surgeons, 1990). The proportion of muscle fibre types appears to be genetically determined and each fibre carries its own specific type of innervation. In general slow muscles are described as 'red', and fast as 'white' because of their paler colour. Fast fibres rely predominantly on glycolytic respiration which is ideally suited to the calf muscles of the sprinter, whereas slow fibres have a predominantly aerobic metabolism.

Muscle soreness after strenuous eccentric exercise is consistent with ultrastructural damage. Muscle fibres hypertrophy in response to training, thus increasing their cross-sectional area and consequently, available force of contraction. Type I (slow) fibres respond to endurance training and type

Table 4.1 Classification of muscle fibre type

Type		*Response*	*'Colour'*	*Function*	*Metabolism*
I		Slow	Red	Endurance	Aerobic
II	A	Fast	White	Power, speed	Glycolytic
	B				
	C				
	M				

II (fast) to resistance exercises. Muscle weakness and fatigue predispose muscle to injury probably because of diminished ability to resist muscle stretch by active contraction. Animal studies show that muscle can absorb more energy if it has higher active force production while being stretched. Hence appropriate, scheduled training and warm-up before events are important.

Muscle injury may occur by indirect or direct means.

Indirect muscle injury

Indirect or strain injuries commonly arise at the myotendinous junction and are frequently associated with sports. An inflammatory response ensues, manifest clinically by pain, swelling and tenderness. As this inflammatory reaction settles, the area of damage is replaced by fibrosis. In most cases a full recovery is achieved.

If injured muscle is to be immobilized, the lengthened position should be adopted to prevent muscle shortening and possibly a greater chance of subsequent injury. Reduced force of muscle contraction after injury may be due to neural inhibition as well as muscle damage. Hence electrical stimulation may have a part to play in maintenance of bulk and strength of an immobilized muscle (American Academy of Orthopaedic Surgeons, 1990).

Direct muscle injury

Direct injury to muscle may be of traumatic or ischaemic origin. Trauma may be crushing or lacerating in nature. Ischaemia will arise from interruption of blood supply or raised pressure within a muscle compartment, both situations requiring immediate surgical intervention. Ischaemia of over 6 hours is generally held to produce irreversible damage within the muscle. Ultrastructural studies of muscles exposed to 3 hours of tourniquet ischaemia have shown that electron microscopic changes are evident within this period. Five hours of tourniquet ischaemia will produce irreversible damage in the animal model (Patterson and Klenerman, 1981). The nature of direct lacerating or crushing injury, perhaps associated with a fracture, is different but, if the muscle is devitalized, the end result will be the same as for ischaemia, namely muscle necrosis. When this occurs an inflammatory response is initiated followed by intensive phagocytosis to remove the dead muscle cells. Myoblasts may then proliferate to form new cells, although the scope for this in man is limited. Large areas of dead muscle are therefore replaced by fibrous scar tissue. Of critical importance is the involvement of nerve by the injury. Denervation will lead to atrophy of the involved segment of the muscle. Despite such loss of muscle mass, power of contraction may be restored by hypertrophy of the remainder, consequent to regular use and training of the muscle.

In the clinical situation, the primary aim is to minimize the muscle loss. This is achieved by prompt surgical wound care, reduction of associated fractures, release of raised intracompartmental pressure and revascularization where necessary.

Nerve injury

The scientific and clinical knowledge of nerve injury is based on the works of the Medical Research Council (1954), Woodhall and Beebe (1956), Seddon (1975) and Sunderland (1978), much of which was accumulated at the Royal National Orthopaedic Hospital, Middlesex, from the experience of World War II. Although a vast amount of microanatomical and physiological knowledge has been gathered in the last 50 years, there is no convincing evidence that the outcome of nerve injury has improved within the same period. This is related to the staggering complexity of the nervous system and also perhaps to excessive expectations placed on its regenerative capacity. Considering that the peripheral nervous system develops at week six of gestation, growing in accord with the spine and limb buds, it is quite remarkable that regeneration of injured neurones back to their original receptors may be expected in the mature adult. The principle of the response to nerve injury is that of individual cellular regeneration as opposed to the process of repair which takes place in tissues such as solid organs or bone. If results of nerve repair are to improve, advances in research must be geared to the cellular and molecular level. Many apparent advances in surgical technique have made virtually no difference to results.

From the clinical point of view the essential points in management are:

1 Early diagnosis of nerve injury.
2 The immediate identification of those injuries which require intervention, in other words, the determination of the degenerative lesion.

Microanatomy and physiology

The neurone consists of the cell body, associated dendrites and usually one axon which may be as long as one metre. Large numbers of axons are gathered in fascicles along with their own microcirculation and the supportive tissues of endoneurium, perineurium and epineurium to form the nerve trunk (Fig. 4.2). All nerve fibres are surrounded by Schwann cells. Myelinated fibres are sheathed by individual Schwann cells in multiple layers, whereas groups of unmyelinated fibres share the enveloping Schwann cell.

The continuity of the myelin sheath is interrupted by the nodes of Ranvier where the Schwann cells interdigitate. The Schwann cell plasma membrane in combination with endoneurial reticular and collagen fibres forms the 'endoneurial tube'. Erlanger and Gasser in 1937 defined that the conduction velocity is proportional to the diameter of the myelinated fibre (Lundborg, 1988).

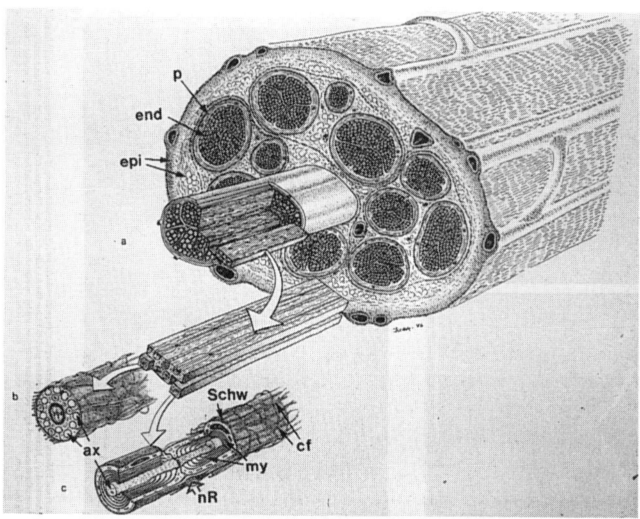

Figure 4.2 Microanatomy of a peripheral nerve trunk and its components. (a) Fascicles surrounded by a multilaminated perineurium (p) are embedded in loose connective tissue, the epineurium (epi). The outer layers of epineurium are condensed into a sheath. (b) and (c) illustrate the appearance of unmyelinated and myelinated fibres respectively. Schw: Schwann cell; my: myelin sheath; ax: axon; nR: node of Ranvier. (Reproduced from Lundborg, G. 1988, Nerve and Injury Repair by kind permission of Churchill Livingstone, London.)

Connective tissue

The perineurium plays a vital role as a physical and diffusion barrier which protects the nerve fibres from direct trauma, ischaemia and toxic substances. The microvascular system supplies metabolic needs by means of parallel, longitudinal vessels which intercommunicate at all levels within the nerve (Fig. 4.3).

Profuse collaterals and multidirectional flow preserve the blood supply, for example, when a nerve is mobilized. There is much concern among surgeons about the risks of mobilizing nerves and resultant precipitation of nerve palsy by ischaemia. Studies in the rabbit demonstrate that the entire sciatic nerve can be mobilized without compromise to nerve function (Lundborg, 1988). It is known that intraneural vessels have their own sympathetic nerve supply which may play a part in the development of reflex sympathetic dystrophy following immobilization or injury.

Axonal transport

In view of the considerable length of many axons, there is a necessity for transport systems to nourish the neurone. This transport occurs in antegrade, that is proximal to distal, and retrograde directions. Antegrade transport provides structural materials and transmitter substances, whereas retrograde transport is thought to dispose of 'waste' products and also to bear neurotrophic factors in the proximal direction. Such factors, although poorly defined at present, play a significant part in nerve function and also in nerve regeneration. With regard to the former, there is now good

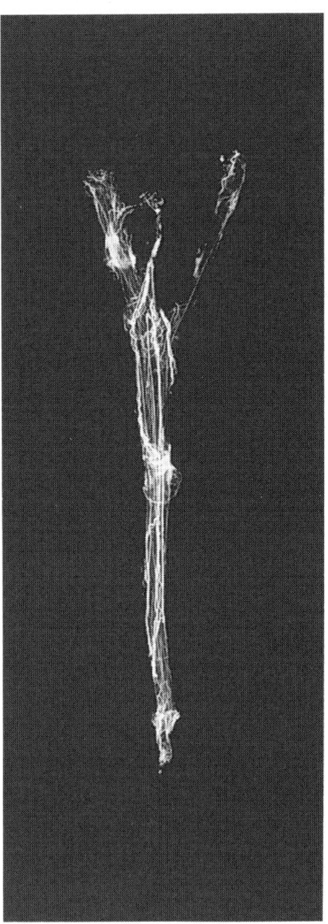

Figure 4.3 Radiographic perfusion study of rabbit sciatic nerve demonstrating multiple intercommunicating longitudinal vessels (with kind permission of Mr C.J. McCullough).

evidence (Dahlin and Lundborg, 1990) that compression of one segment of a neurone renders another region of the neurone, whether proximal or distal, more susceptible to similar compression. This phenomenon is described as the 'double crush syndrome', the commonest example being the simultaneous occurrence of carpal tunnel syndrome and cervical root compression with greater frequency than would be expected by coincidence. The increased susceptibility to simultaneous pressure at a separate site appears to be mediated by limitation of retrograde flow in the axon which leads to degenerative changes within the central cell body. This is a relatively recent finding because the study of pressure injury to nerve has largely focused on the region of the nerve under pressure rather than on the cell body. Following division of the nerve, the absence of factors transported in a retrograde direction may be responsible for initiation of axonal sprouting (Varon and Williams, 1986).

Types of injury and the degenerative lesion

Seddon, in 1943, considerably simplified matters by describing three types of nerve injury (Table 4.2). He coined the term neurapraxia which implies a transient conduction block without axonal interruption, secondary to compression or traction injury. This is a dangerous clinical diagnosis which must be made with caution and which should not be used when there is a related open injury or surgical incision. Axonotmesis describes axonal interruption with preservation of the endoneurial tube and hence the opportunity for spontaneous reinnervation of terminal receptors. Both these lesions will recover spontaneously, the former in hours to weeks, and the latter in months, although with less likelihood of full recovery.

Table 4.2 Classification of nerve injury

	Neurotmesis	*Axonotmesis*	*Neurapraxia*
Pathological			
Anatomical continuity	May be lost	Preserved	Preserved
Essential damage	Complete disorganization	Nerve fibres interrupted; Schwann sheaths preserved	Selective demyelination of larger fibres: no degeneration of axons
Clinical			
Motor paralysis	Complete	Complete	Complete
Muscle atrophy	Progressive	Progressive	Very little
Sensory paralysis	Complete	Complete	Usually much sparing
Autonomic paralysis	Complete	Complete	Usually much sparing
Electrical phenomena			
Reaction of degeneration	Present	Present	Absent
Nerve conduction distal to the lesion	Absent	Absent	Preserved
Motor-unit action potentials	Absent	Absent	Absent
Fibrillation	Present	Present	Occasionally detectable
Recovery			
Surgical repair	Essential	Not necessary	Not necessary
Rate of recovery	1–2 mm a day after repair	1–2 mm a day	Rapid: days or weeks
March of recovery	According to order of innervation	According to order of innervation	No order
Quality	Always imperfect	Perfect	Perfect

From H.J. Seddon (1975) with kind permission of Churchill Livingstone.

The key to clinical practice is differentiating the above two entities from the degenerative lesion or neurotmesis, when the axons and their endoneurial tubes are transected or injured in continuity by a traction lesion. Clinical examination after nerve injury is unreliable because of anomalous muscle innervation, supplementary action, sensory overlap and the possibility of incomplete division. Function may appear minimally diminished even when the majority of fibres have been cut. For these reasons surgical exploration is the only sure means of diagnosis.

The brachial plexus

The differentiation of preganglionic and postganglionic brachial plexus lesions is difficult but important because of the differences in treatment and prognosis. In preganglionic lesions the nerve root is avulsed from the spinal cord proximal to the posterior root ganglion, thus preserving the axon reflex which is the basis for a positive histamine test (Bonney, 1954). Electrophysiological investigations are of little help in the early phase as action potentials distal to the point of nerve division may be recorded for up to 3 weeks after injury until demyelination is complete. Computed tomography combined with myelography (Marshall and DeSilva, 1986), will demonstrate a meningocele at the site of root avulsion, particularly at C5 and C6 levels although this investigation has a false positive rate of about 10%. Poor prognostic signs suggestive of root avulsion are: complete paralysis of the arm, anaesthesia of arm and sensory loss extending above clavicle, burning pain, Horner syndrome and associated vascular injury (Table 4.3).

Table 4.3 Brachial plexus injury – poor prognostic signs

1 Complete paralysis of arm and thoracoscapular muscles
2 Anaesthesia of limb and sensory loss above clavicle
3 Burning pain in anaesthetic hand
4 Horner syndrome
5 Associated vascular injury

Postganglionic lesions may be supraclavicular, infraclavicular or in exceptional circumstances both. The infraclavicular injury may be of one nerve, such as the axillary nerve, or of a combination of nerves. Infraclavicular injuries are of particular concern as there is a 50% incidence of both associated vascular injury and concurrent neighbouring fracture.

Brachial plexus injuries are rarely clean-cut and are usually of the traction type, which may make definition of site and length of lesion difficult.

Degeneration

After injury, well-defined proximal and distal changes occur in the nerve cell (Figure 4.4).

In the proximal segment the nerve cell body enlarges, the nucleus becomes peripheral and chromatolysis or loss of Nissl granules is noted. RNA and protein content increase in preparation for axonal regeneration. The carriage by the axonal transport system increases. As stated previously the chromatolysis may be secondary to loss of nerve growth factor from retrograde axonal transport. Retrograde regeneration occurs in the axon for a variable distance depending on the severity of the lesion.

Distal to injury, Wallerian degeneration ensues consisting of lysis of axoplasm and fragmentation of myelin sheaths leaving a collapsed endoneurial tube containing Schwann cells. This process may be triggered by calcium.

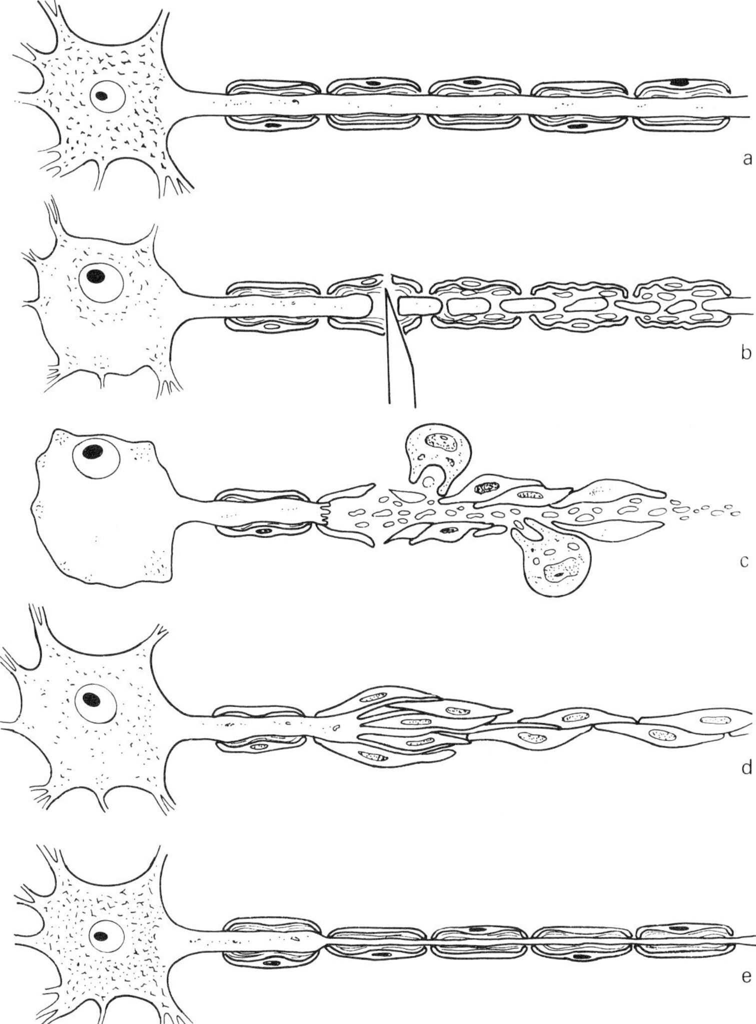

Figure 4.4 Degeneration and regeneration of myelinated fibre. (a) Normal appearance. (b) Transection of the fibre results in distal fragmentation of axon and myelin. In the proximal segment degeneration occurs at least to the nearest node of Ranvier. (c) In the distal segment Schwann cells proliferate. Macrophages and Schwann cells phagocytose debris material. (d) The Schwann cells in the distal segment have lined up in bands of Bungner. Sprouting occurs from the cut axonal stump. Advancing sprouts are embedded in Schwann cell cytoplasm. (e) 'Axonal' connection with periphery, maturation of nerve fibre. Sprouts which do not link up with the periphery may atrophy and disappear. The cell body response during these phases includes swelling, migration of the nucleus to the periphery and condensation of basophilic material (chromatolysis). (Reproduced from Lundborg, G. 1988, *Nerve and Injury Repair* by kind permission of Churchill Livingstone, London.)

Regeneration

In the proximal nerve segment lines of endoneurial collagen are then laid down and Schwann cells begin to proliferate rapidly in columns known as bands of Bungner from 1 to 5 days after division. The Schwann cells manufacture myelin and also probably neurotrophic factors such as laminin. Within a few days myelinated axons produce terminal and collateral sprouts which advance down the endoneurial tubes preceded and surrounded by Schwann cells. Regenerating units aggregate into minifascicles particularly at the periphery of the proximal stump. This phenomenon is known as compartmentation and may represent an attempt to recreate the perineurial barrier. Assuming nerve continuity has been restored, large numbers of sprouts penetrate the Schwann cell columns. The axonal sprouts migrate down the endoneurial tubes ultimately reaching receptors. It is obvious that this process is haphazard and therefore the matching of appropriate axonal sprouts and receptors is minimal. Those sprouts that reach distal receptors will mature and enlarge, whereas those that do not will atrophy. Histological studies in animals have demonstrated that the fibre count within regenerating nerve or nerve graft may be artificially high in the early stages because of axonal proliferation in competition for receptors. Transverse sections made many months after repair will tend to show fewer numbers of myelinated axons. Each regenerating neurone appears to carry its own code for myelination consistent with that prior to injury and may accordingly switch on a Schwann cell to produce myelin.

Nerve growth factors

The initiation of axonal regeneration and the traverse of the site of injury is certainly under the control of nerve growth factors, but the nature and origin of these factors is still unclear. Rather like the race between bone healing and impending failure of internal fixation, there is a similar contest between axonal regeneration and distal receptor degeneration. This may partly account for the poor results of proximal nerve injuries. Regeneration proceeds at 2–3.5 mm per day in rats and 1–2 mm per day in humans. Following regeneration the quality of nerve fibres measured in terms of axon diameter and myelin sheath thickness does not return to normal.

Factors influencing axonal growth

Extensive microenvironmental experiments using nerve regeneration chambers have shown that for axonal growth a distal nerve segment is necessary (Varon and Williams,1986; Lundborg 1988). Schwann cells may offer an inductive cellular matrix incorporating laminin and fibronectin which appear to be secreted by fibroblasts or endothelial cells in the nerve stumps. A variety of neuronotrophic factors exist in experimental nerve chamber fluid, each having a specific effect on sensory, sympathetic and spinal cord neurones. The main source for these factors is probably the Schwann cells. In addition, factors in the fluid appear to promote Schwann cell adhesion, motility and proliferation, thus in theory, improving the environment for regenerating axons.

Various attempts have been made to influence the environment at the site of regeneration. The systemic administration of various substances such as thyroid hormone, ACTH, ganglioside and cycic AMP has been shown to stimulate axonal regeneration in some animal experiments (Lundborg, 1988).Raji and Bowden (1983) demonstrated that pulsed electromagnetic fields improve nerve regeneration and indeed limb function in rats following crush injury to the sciatic nerve. The results of topical and systemic administration of scar inhibitors upon fibrosis at the nerve join and consequently regeneration appear inconclusive (Lundborg, 1988).

Surgical repair

Factors which influence the outcome of nerve repair include age, level of injury, the type of nerve involved (mixed or unmixed), the nature and force of injury and the timing and technique of repair. Children have a markedly better prognosis than adults, probably because of their regenerative capacity, their relearning ability and also the shorter distances of regeneration required. Gilbert (1991) and Narakas (1987) have demonstrated dramatic results from repair of obstetric brachial plexus lesions. Míxed nerves such as the ulnar and sciatic offer poor prospects, particulary if the lesion is proximal. Long segments of damage, for example in traction or gunshot lesions will not do well.

Clinical studies suggest that early repair is beneficial so long as the procedure is performed by an experienced surgeon under favourable circumstances, preferably with magnification (Table 4.4). Some injuries are associated with major soft tissue trauma and vascular damage. These lesions should be dealt with immediately, as soon as the general condition of the patient permits, with careful attention to the viability of blood vessels, muscle and subsequently nerve. Nerve repair requires the presence of a healthy tissue bed and the absence of contamination or infection. Clinical evidence suggests that repair of associated arterial injury, for example, when both ulnar nerve and artery are divided, leads to a better functional outcome (Leclercq *et al.*, 1985). It is tempting to think that this is because of improved vascularization of the nerve, but the underlying reason is probably that the arterial repair relieves tension upon the nerve anastomosis. This theory is supported by the fact that many such arterial anastomoses thrombose soon after operation. Millesi (1977) and subsequently Terzis (1987) have demonstrated clearly that tension at the nerve suture site interferes with the functional result. It is often possible to bring nerve ends together by mobilization and flexion of the neighbouring joint. Should this not be possible without tension, then recourse must be made to nerve graft.

With the advent of the operating microscope, much was expected in terms of improved results using fascicular repair of major trunk nerves. Synthetic, inert microsutures are essential in order to minimize the trauma of insertion and limit consequent fibrosis and scarring. Recent experience with fibrin glue suggests that major nerve repair, for example in the brachial plexus, is as good or possibly even better when fibrin glue is used instead of microsutures. In addition, a considerable amount of operating time is saved.

Table 4.4 Surgical requirements for successful nerve repair

1 Experienced surgeon with adequate facilities
2 Minimal delay
3 Skeletal stability
4 Healthy, clean tissue bed for repair
5 Lack of tension at join
6 Minimal trauma of handling and suture
7 Magnification preferable

Grafts for repair

Until recently the only clinically recognized source of graft for bridging large defects in nerve has been autogenous, mainly from cutaneous nerves of the forearm and leg. The harvesting of these grafts confers its own morbidity in the form of scar and further sensory loss. Such grafts are sectioned into appropriate lengths and used as multiple cables to bridge truncal defects, for example in the median nerve or the supraclavicular brachial plexus. Sunderland (1978) demonstrated the complex interweaving nature of nerve trunks with his sequential topographical studies. The use of cable grafts makes restoration of remotely normal topography impossible. Under special circumstances the ulnar nerve may be used as a large trunk graft, preferably in its vascularized form, for example, when bridging an upper trunk defect of the brachial plexus in the presence of avulsion of the lower roots (Birch *et al.*, 1988). In this situation it is acceptable to sacrifice the ulnar nerve.

An obvious source of graft would appear to be the human cadaver and this was tried by Seddon (1975) with no success whatsoever. More recent studies by Bain *et al.* (1988) in the laboratory rat appear to show that cyclosporin A immunosuppression allows regeneration through a rat sciatic nerve allograft equal to that through a similarly placed isograft. The experience of the author in this field using shorter allografts suggests that cyclosporin A has no effect on functional outcome or nerve morphology in the rat. The idea of a nerve allograft bank is appealing but unlikely to become reality because of the toxicity of immunosuppression and the poor vascularization of large trunk grafts.

Recent laboratory and clinical studies have demonstrated that muscle autograft, denatured by deep freezing then rapid thawing in distilled water may be used to bridge peripheral nerve defects. This technique has been applied to rats (Glasby *et al.*, 1986a), primates (Glasby *et al.*, 1986b) and sheep (Glasby *et al.*, 1990) with success and has been shown to work in the clinical situation (Norris *et al*, 1988).

If orientated longitudinally the denatured muscle graft provides a sarcolemmal network which conducts regenerating axons in the same manner as a nerve graft. Particular applications for this technique may be the bridging of truncal defects where adequate nerve graft cannot be acquired and also for restoration of nerve continuity for painful neuroma following injury (Stirrat, Birch and Glasby, 1991). The maximum feasible length of a muscle graft is about 6 cm. Nerve regeneration has been achieved in animals through pseudosynovial sheaths (Lundborg, 1988) and many synthetic alternatives have been tried.

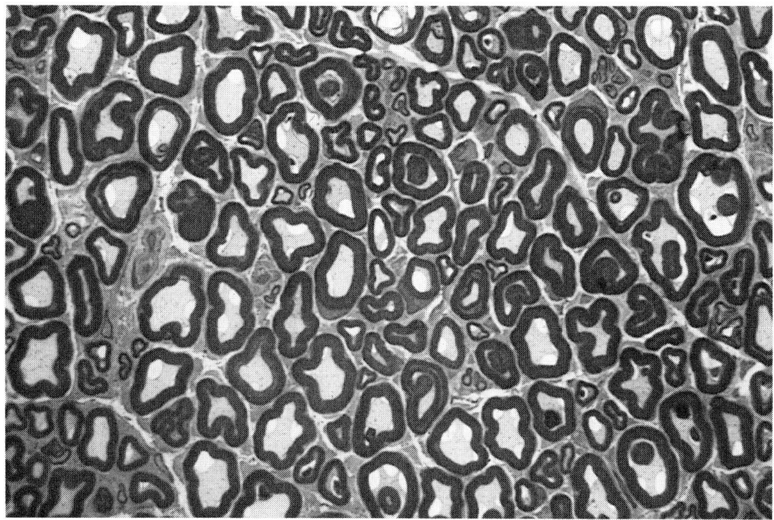

Figure 4.5 Cross-section of sheep sciatic nerve distal to interposed muscle graft showing regenerated myelinated nerve fibres. (Toluidine blue; × 1000.) (Reproduced with kind permission of M.A. Glasby.)

Nerve transfer for the brachial plexus injury

The management of avulsion lesions of the brachial plexus requires special consideration. Early exploration within 6 weeks is important to determine the extent of the lesion, clarify the prognosis and to attempt reconstruction before excessive scarring surrounds the plexus. The primary aim is to restore shoulder and elbow control, particularly if hand function remains. A prime consideration is the motivation and attitude of the patient, which will govern the advantage taken of returning function.

A revolutionary development was applied by Sir Herbert Seddon (1963) in the form of nerve transfer (Table 4.5). This has now become established in Europe and the UK as a method of treatment which offers restoration of function in the presence of preganglionic avulsion of nerve roots. The accessory nerve may be transferred to the suprascapular nerve in expectation of restoring active shoulder abduction of MRC power grade 4 at 6 months (Table 4.6). There is some evidence to suggest that the presence of a good hand predisposes to a better outcome of accessory nerve transfer (Stirrat , Patterson and Birch, 1991 unpublished data). The intercostal nerves may be transferred to the musculocutaneous nerve and median nerve origin, to provide elbow flexion and protective sensation in the hand, respectively (Narakas and Hentz, 1988). Elbow flexion of power grade three or four may be anticipated after a delay of 12 months. Restoration of function by intercostal nerves is remarkable and presumably occurs as a result of cortical relearning. Certainly as regeneration proceeds, the patient is initially aware of movement in the reinnervated biceps during respiration. With practice the patient appears to be able to suppress this.

Brachial plexus reconstruction is a major specialized procedure and results may not be evident for 2 years. The essential principles are immediate

Table 4.5 Nerve transfers for avulsion of C5 and C6 roots

Transfer	*Anticipated result*	*Approximate time to recovery (months)*
Accessory to suprascapular	Grade 4 abduction of shoulder to 60°	6
Intercostal to musculocutaneous	Grade 3 or 4 biceps	12
Intercostal to origin of median	Protective sensation in hand and pain relief	12

Table 4.6 Medical research council grading of motor power

0	Complete paralysis
1	Muscle responds with slight flicker on volition
2	Can move the part on volition when gravity eliminated
3	Can move the part only against gravity
4	Can move the part against gravity and some resistance
5	Normal power

treatment of associated vascular and bone injuries, exploration within 6 weeks for diagnosis and treatment so that the patient may be informed of the prognosis and prompt referral to a rehabilitation unit so that the patient may return to work.

Assessment of nerve regeneration

In the laboratory this can be done in great detail using animals which only partially relate to the situation in the human. Investigation may be by functional, electrophysiological, or histological means. Functional assessment has been taken as far as analysis of rat gait following nerve repair (de Medinacelli, Freed and Wyatt, 1982). More traditional methods include regular testing of withdrawal reflexes and post-mortem weighing of muscles.

Electrophysiological studies may be performed accurately on exposed nerves at anaesthesia. Compound nerve action potentials will demonstrate nerve function, although they are not accurate in a quantitive sense. Nerve histology provides information regarding number of fibres, fibre diameter and myelin sheath thickness. The number of fibres in a regenerating nerve graft is of limited relevance as a high proportion will never reach receptors and will ultimately atrophy. Fibre diameter and myelin sheath thickness may give an estimate of the quality of regeneration. Horseradish peroxidase injection into the distal nerve is a useful technique that allows confirmation of regeneration by demonstrating continuity between spinal cord and distal receptor.

The assessment of nerve injury in the human is almost entirely clinical. Functional testing as described by Moberg (1970) and Dellon and Kallman (1983) has superseded the basic testing of motor power and sensation by MRC grading (see Table 4.6). The Tinel sign gives a rough guide to the progress of nerve regeneration and is the clinical manifestation of the advancing wave of regenerating sensory axons.

Rehabilitation

Perhaps the greatest problem for the patient is the pain associated with some nerve injuries, particularly those of the brachial plexus and lesions secondary to missile wounds. Unremitting pain and hypersensitivity may further limit limb function and precipitate depression and despair. This vicious circle is difficult to break. Surgical priorities for pain control are early treatment, restoration of nerve continuity, removal of compressive or irritative factors and nerve transfer for preganglionic lesions. Sympathectomy, stellate ganglion block and guanethidine blocks may be employed depending on the site of injury. Medication with carbamazepine or antidepressants, transcutaneous nerve stimulation and desensitization may help to break the pain cycle. Perhaps the most important therapy is early return to gainful employment after appropriate rehabilitation.

Prospects for the future

At present research interest focuses on identification and isolation of nerve growth factors, extracellular matrix molecules and regeneration of neurites in the central nervous system (Carlstedt *et al.*, 1987; Fehlings and Tator, 1988). It has been shown that oligodendrocytes inhibit CNS regeneration and that their ablation in the neonatal rat allows CNS neurones to regenerate in the spinal cord (Bandtlow, Zachleder and Schwab, 1990). Such findings offer exciting prospects for the future.

Conclusion

A vast amount of laboratory and clinical research has been performed over the last 50 years in the field of peripheral nerve injury and repair. With the exception of the advent of nerve transfer, clinical results in expert hands are essentially unchanged. However, it is vital that all clinicians understand the importance of early diagnosis and prompt management of nerve injury. This offers the patient the best possible chance of complete rehabilitation.

Acknowledgements

I would like to thank Mr R Birch and Dr C Green for their invaluable teaching, advice and support. I greatly appreciate the assistance of Mrs C. Dow in the collation and typing of the manuscript.

References

American Academy of Orthopaedic Surgeons (1990) *Orthopaedic Knowledge Update*, **3**, 93–98.

Bain, J.R., Mackinnon, S.E., Hudson, A.R., Falk, R.E., Falk, J.A. and Hunter, D.A. (1988) The peripheral nerve allograft: an assessment of regeneration across nerve allografts in rats immunosuppressed with Cyclosporin A. *Plastic and Reconstructive Surgery,* **82,** 1052–63.

Bantlow, C., Zachleder, T. and Schwab, M.E. (1990) Oligodendrocytes arrest neurite growth by contact inhibition. *J. Neuroscience,* **10,** 3837–48.

Birch, R., Bonney, G., Dowell, J. and Hollingdale, J. (1991) Iatrogenic injuries of peripheral nerves. *Journal of Bone and Joint Surgery,* **73B,** 280–82.

Birch, R., Dunkerton, M., Bonney, G. and Jamieson, A.M. (1988) Experience with free vascularised ulnar nerve graft in repair of infraclavicular lesions of the brachial plexus. *Clinical Orthopedics,* **237,** 96–104.

Bonney, G. (1954) The value of axon responses in determining the site of lesion in traction injuries of the brachial plexus. *Brain,* **77,** 588–609.

Bonney, G. (1983) Peripheral nerve lesions: management. In: (Harris N.H., ed.) *Postgraduate Textbook of Clinical Orthopaedics.* Bristol: Wright, 719–30.

Bonney, G. (1986) Iatrogenic injuries of nerves. *Journal of Bone and Joint Surgery,* **68B,** 9–13.

Carlstedt, T., Dalsgaard, C.J., Molander, C. (1987) Regrowth of lesioned dorsal root nerve fibres into the spinal cord of neonatal rats. *Neuroscience Letters,* **74** (1), 14–8.

Dahlin, L.B. and Lundborg, G. (1990) The neurone and its response to peripheral nerve compression. *Journal of Hand Surgery,* **15B** 5-10.

Dellon, A.L. and Kallman, C.H. (1983) Evaluation of functional sensation in the hand. *Journal of Hand Surgery,* **8A,** 865–870.

de Medinacelli, L., Freed, W. and Wyatt, R.J. (1982) An index of the functional condition of rat sciatic nerve based on measurements made from walking tracks. *Experimental Neurology,* **77,** 634–643.

Fehlings, M.G. and Tator, C.H. (1988) A review of spinal cord regeneration. *Advances in Trauma,* **3,** 223–240.

Gattuso, J.M., Davies, A.H., Glasby, M.A., Gschmeissner, S.E. and Huang, C.L.-H. (1988) Peripheral nerve repair using muscle autografts. *Journal of Bone and Joint Surgery,* **70B,** 524–9.

Gilbert, A. and Whitaker, I. (1991) Obstetrical brachial plexus lesions. *J. Hand Surgery,* **16B,** 489–91.

Glasby, M.A., Gschmeissner, S., Hitchcock, R.J.I. and Huang, C.L-H. (1986a) Regeneration of the sciatic nerve in rats. *Journal of Bone and Joint Surgery,* **68B,** 829–33.

Glasby, M.A., Gschmeissner, S.E., Huang, C.L-H. and De Souza, B.A. (1986b) Degenerated muscle grafts for peripheral nerve repair in primates. *Journal of Hand Surgery,* **11B,** 347–51.

Glasby, M.A., Gilmour, J.A., Gschmeissner, S.E., Hems, T.E.J. and Myles, L.M. (1990) The repair of large peripheral nerves using skeletal muscle autografts: a comparison with cable grafts in the sheep femoral nerve. *British Journal of Plastic Surgery,* **43,** 169–78.

Leclercq, D.C., Carlier, A.J., Khuc, T., Depierreux, L. and Lejueune, G. (1985) Improvement in the results of sixty-four ulnar nerve sections associated with arterial repair. *Journal of Hand Surgery,* **10A,** 997–9.

Lundborg, G. (1988) *Nerve Injury and Repair.* Churchill Livingstone, Edinburgh.

Marsh, D. and Barton, N. (1987) Does the use of the operating microscope improve the results of peripheral nerve suture? *Journal of Bone and Joint Surgery,* **69B,** 625–30.

Marshall, R.W. and DeSilva, R.D.D. (1986) Computerised axial tomography in traction injuries of the brachial plexus. *Journal of Bone and Joint Surgery,* **68B,** 734–38.

Medical Research Council (1954) Peripheral nerve injuries. Spec. Rep. Ser. Med. Res. Coun. No. 282. London, HMSO.

Millesi, H. (1977) Healing of nerves. *Clinical Plastic Surgery,* **4,** 459–73.

Moberg, E. (1976) Sensibility in reconstructive limb surgery. In: (Fredericks, S. and Brody, G.S. eds.) *Symposium on the Neurologic Aspects of Plastic Surgery.* C.V. Mosby, St Louis.

Narakas, A.O. (1987) Obstetric brachial plexus injuries. In: (Lamb, D.W. (ed.) *The Paralysed Hand.* 117–135 Churchill Livingstone, Edinburgh.

Narakas, A.O. and Hentz, V.R. (1988) Neurotisation in brachial plexus injuries. *Clinical Orthopedics,* **237,** 43–56.

Norris, R.W., Glasby, M.A., Gattuso, J.M. and Bowden, R.E.M. (1988) Peripheral nerve repair in humans using muscle autografts. *Journal of Bone and Joint Surgery,* **70B,** 530–33.

Patterson, S. and Klenerman, L. (1981) The effect of pneumatic tourniquet on skeletal muscle physiology. *Acta Orthopaedica Scandinavica,* **52,** 171–75.

Raji, A.R.M. and Bowden, R.E.M. (1983) Effects of high-peak pulsed electromagnetic field on the degeneration and regeneration of the common peroneal nerve in rats. *Journal of Bone and Joint Surgery,* **65B,** 478–92.

Seddon, H.J. (1963) Nerve grafting. *Journal of Bone and Joint Surgery,* **45B,** 447–61.

Seddon, H. (1975) *Surgical Disorders of the Peripheral Nerves.* Edinburgh: Churchill Livingstone.

Stirrat, A.N., Birch, R. and Glasby, M.A. (1991) Applications of muscle autograft in peripheral nerve injury. *Journal of Bone and Joint Surgery,* **73B,** Suppl. II, 166–67.

Sunderland, S. (1978) *Nerve and Nerve Injuries.* Edinburgh: Churchill Livingstone.

Terzis, J. (1987) *Microreconstruction of Nerve Injuries.* W.B. Saunders Co., Philadelphia.

Varon, S. and Williams, L.R., (1986) Peripheral nerve regeneration in the silicone model chamber: cellular and molecular aspects. *Peripheral Nerve Repair and Regeneration,* **1,** 9–25.

Woodhall, B. and Beebe, G.W. (1956) Peripheral nerve regeneration. Veterans Administration Monograph. Washington D.C.: United States Government Printing Office.

Chapter 5

Biomechanics of the hip

David Pring

Anatomy

The hip joint is a ball and socket joint formed by the head of the femur and the acetabulum of the pelvis. This arrangement permits considerable freedom of movement. Both the articular surface of the femur and acetabulum are lined with hyaline cartilage. The articular cartilage of the head is thin peripherally and thicker centrally, whereas the acetabular cartilage is thickest peripherally. The femoral head and acetabular socket are not quite congruent, the femoral head being slightly larger than the acetabulum. Consequently, with small loads the femoral head is only in contact with the periphery of the acetabulum and it is not until greater loads are applied that the femoral head articulates with the floor of the acetabulum.

The hip joint is surrounded by a loose capsule that is thickest anteriorly (the ilio-femoral ligament of Bigelow). With extension this ligament is tightened producing a close-pack position with maximum congruency of the articular surfaces. Hip joint movements are controlled by the surrounding muscles that act as dynamic stabilizers. Muscular contraction is responsible for the majority of the forces that are applied across the hip joint.

The joint reaction force

The joint reaction force is the sum of all forces that are applied to a joint and have to be distributed by that joint. The centre of the femoral head may be regarded as the fulcrum of a lever system. In two-legged stance the joint reaction force of each hip joint is easily calculated, assuming that body weight is shared equally between both hip joints and that there is no muscle activity around the hip. In this situation body weight minus the weight of both legs (in practice 62% of body weight) is shared between each joint (Fig. 5.1).

In practice, however, muscle activity is required to maintain balance and prevent swaying movements. Thus the joint reaction force on each hip during two legged stance is a little greater than one-third body weight.

Denham in his excellent paper on hip mechanics draws the analogy with a flag pole stand (Denham, 1959) (Fig. 5.2).

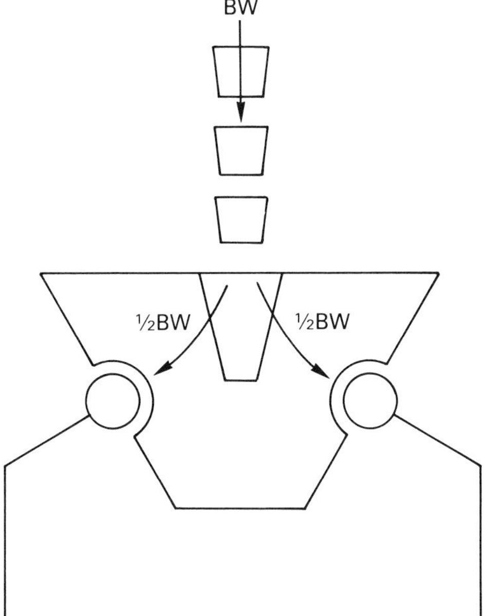

Figure 5.1 Normal two-legged stance. BW = bodyweight minus the weight of both lower limbs

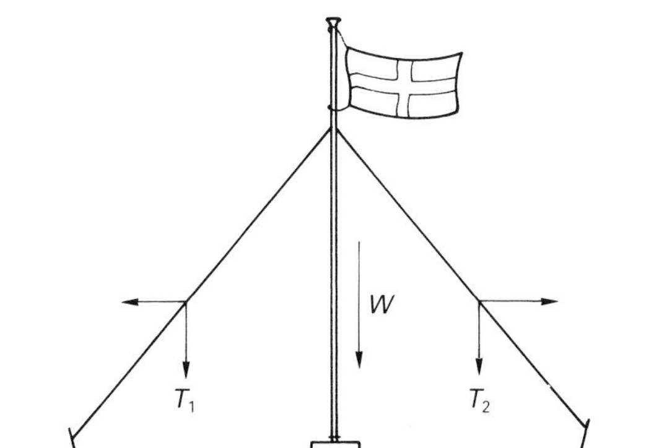

Figure 5.2 Pressure on the flag pole base equals $W + T_1 + T_2$. (Adapted from Denham, 1959 by kind permission of *Journal of Bone and Joint Surgery*.)

The flag pole is maintained in equilibrium by two guy ropes corresponding to the balancing muscles. Body weight is represented by the weight of the flag pole. With the flag pole upright very little tension is required in the guy ropes. What tension there is can be resolved into vertical and horizontal components. Thus the reaction force on the base of the flag pole is $W + T_1 + T_2$.

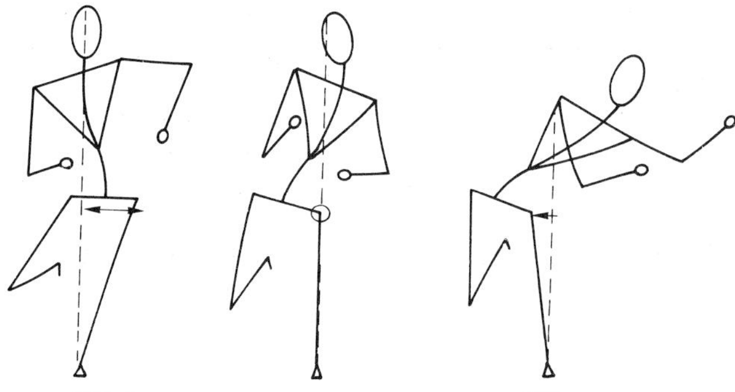

Figure 5.3 Exhibiting the variable relationship between the hip joint and the line of body weight. This is subject to continual change

If, however, the flag pole leans to one side, as in one legged stance, to maintain equilibrium tension in the guy rope is much increased. This is reflected by a marked increase in the reaction force on the base of the flag pole.

In one-legged stance the centre of gravity must shift towards the weight-bearing limb and pass through the weight-bearing foot if equilibrium is to be maintained. If the line of body weight passes through the centre of the hip and the foot, then theoretically, no muscle activity is required (Fig. 5.3).

In this situation the hip joint reaction force would be body weight minus the weight of one lower limb, which represents 81% of body weight. However, when the line of body weight passes medial to the hip joint (Fig. 5.4), (Denham, 1959) an anticlockwise moment is created that has to be resisted by tension in the abductor muscles. This results in an enormous increase in the joint reaction force.

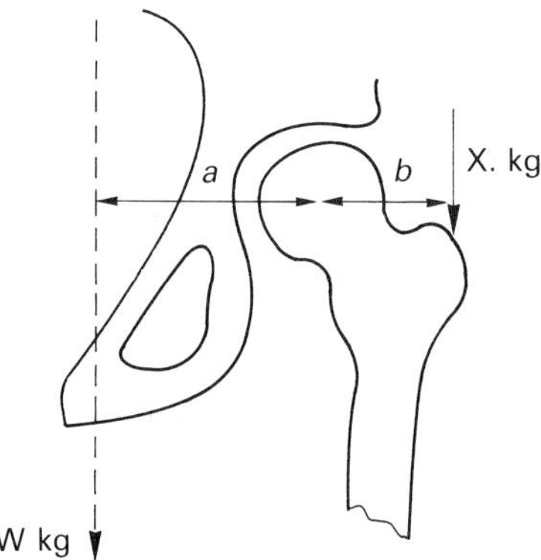

Figure 5.4 The production of an anti-clockwise moment where the line of the body weight passes medial to the hip joint. The lever arm *a* is balanced by the lever arm *b* acted upon by the abductor muscles

The distance between the centre of the head of the femur and the line of body weight is represented by a. This is the lever arm of the body weight and can be varied by a variety of factors including the obliquity of the pelvis, the position of the limbs and the position of the trunk. The anticlockwise turning moment is represented by body weight (minus the weight of one leg) $W \times a$. For equilibrium this is resisted by the abductor force $X \times b$ where b is the distance from the centre of the head of the femur to the insertion of the abductors.

If $W = 70$ Kg, $a = 10$ cm, $b = 5$ cm and $X =$ abductor force

5 cm 10 cm

Δ

X 70 Kg

$70 \times 10 = 5 \times X$
$X = 140$ Kg

The joint reaction force is $W + X$ which is 210 Kg or *three times body weight*. Thus in changing from a two-legged to a one-legged stance the joint reaction force has increased from one third to three times body weight.

How can the joint reaction force be modified?

1. Changing the body weight lever arm

If the centre of gravity is shifted laterally so that the line of body weight passes 5 cm medial to the hip joint. Then for the equilibrium

5 cm 5 cm

Δ

X 70 Kg

$70 \times 5 = 5 \times X$
$X = 70$ Kg

Joint reaction force = 70 + 70 = 140 Kg or two times body weight, so by moving the line of body weight considerable changes occur in the forces applied to the hip joint. Clinically this is exactly what the limping patient does when he throws his body weight over the affected hip to minimise the body weight lever arm or even to reverse it.

2 Change of Body weight

Now let us consider a change in the weight of the patient. W increased from 70 Kg to 80 Kg.

5 cm 10 cm

Δ

X 70 Kg

$80 \times 10 = X \times 5$
$X = 160$ Kg

Joint reaction force = 160 + 80 = 240 Kg. Therefore, for every 1 Kg increase in body weight there is a 3 Kg increase in the joint reaction. Obviously, a reduction in 1 Kg weight is matched by a 3 Kg reduction in the force across the hip joint.

3. Changing the Abductor Lever Arm

Clearly changing b, the moment arm of the abductor muscles will also alter the joint forces. b is related to the femoral neck length, the neck shaft angle and following total joint replacement the offset of the prosthesis. Let us consider foreshortening neck length as may occur following a fracture of the neck of the femur: $b = 2.5$ cm.

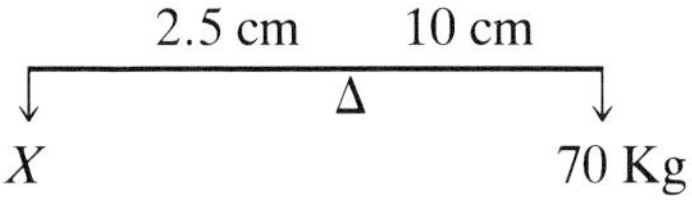

2.5 cm 10 cm

Δ

X 70 Kg

$W \times \mathrm{a} = \mathrm{X} \times \mathrm{B}$
$70 \times 10 = X \times 2.5$
$X = 280$ Kg

Halving the abductor moment arm doubles the required abduction force. The joint reaction force is also considerably increased. $W + X = 70 + 280 = 350$ Kg compared with 210 Kg originally.

The converse, lateralizing the insertion of the abductor muscles increases the abductor moment arm, so reducing the abductor muscle force required and consequently reducing the joint reaction force.

4. Medializing the femoral head

Medialization of the femoral head by deepening the acetabulum during total joint replacement might be expected to reduce the absolute joint forces by reducing the moment arm of the body weight. If the acetabulum is medialized 2.5 cm:

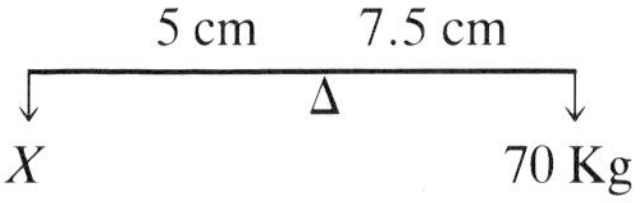

5 cm 7.5 cm

Δ

X 70 Kg

$70 \times 7.5 = X \times 5$
$X = 105$ Kg

Joint reaction force = 105 + 70 = 175 Kg. But this gain is outweighed by other factors which will be discussed later.

5. Medializing the support of the fulcrum

It should be noted that medialization of the femur as occurs after the McMurray osteotomy does not have an effect on the joint reaction force because the insertion of the abductors and their moment arm remains unchanged (Fig. 5.5).

Merely moving the support column under a pivot does not change the mechanical stressing of that pivot. It does however reduce the bending

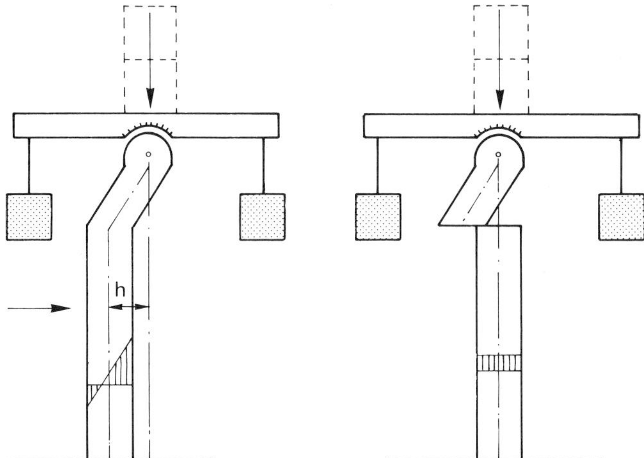

Figure 5.5 Medialization of the femur does not alter the joint reaction force but reduces the lever arm of the force *h* acting on the column

stresses of an offfset column to compressive forces in a vertical column. The biomechanical advantage achieved by the McMurray osteotomy is due to the proximal displacement of the femoral shaft and not by medialization of the femur (Fig. 5.6). Proximal displacement of the femoral shaft results in effectively shortening the ilio-psoas and the adductor muscles. These muscles are therefore weaker leading to a reduction in the joint reaction force.

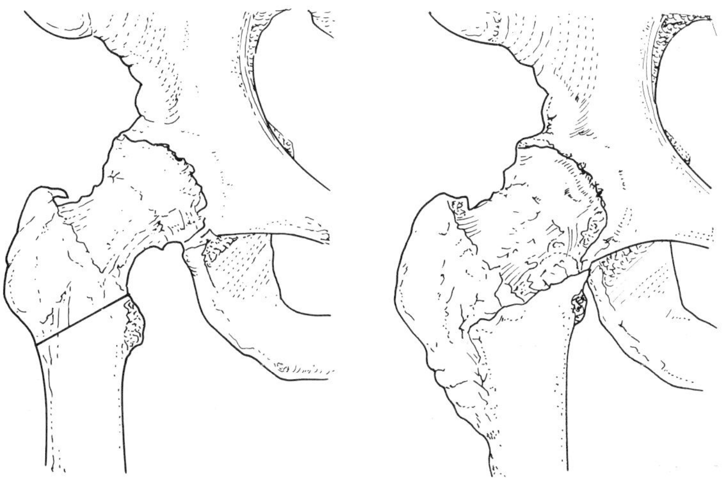

Figure 5.6 McMurray osteotomy

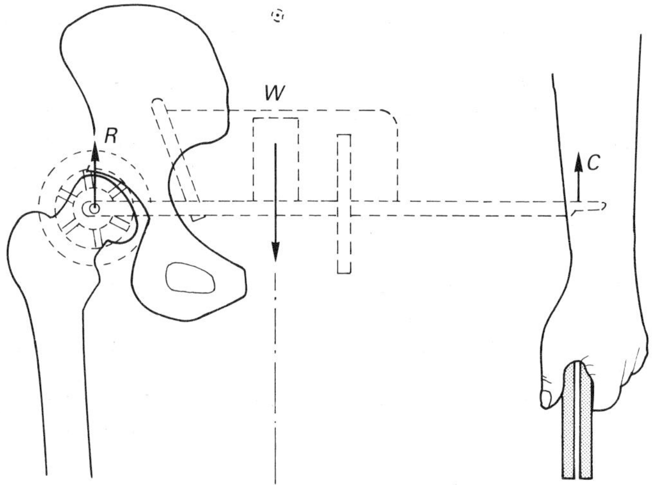

Figure 5.7 The effect of using a walking stick. This is similar to the use of the lever arm on a wheelbarrow. *W*, forces exerted by the body; *R*, the resultant of the forces W and C.

6. The Effect of a Walking Stick

The hip joint reaction force can be considerably reduced by the use of a walking stick (Fig. 5.7).

If the subject holds the walking stick 50 cm from the weight-bearing hip and pushes down with a force of 10 Kg, the reaction force is calculated thus:

10 Kg

5 cm 10 cm 50 cm

↓ Δ ↓

X 70 Kg

$70 \times 10 = (5 \times X) + (10 \times 50)$
$X = 40$

Joint reaction force $= 40 + 70 - 10 = 100$ Kg.

Thus the walking stick has reduced the joint reaction force from 210 Kg to 100 Kg.

(The above calculations for simplicity have omitted a number of factors including the direction of the resultant force of the abductors, the position of the lines of supported and total body weight and the effects of acceleration and deceleration during locomotion. Nevertheless, the above calculations are accurate enough to show the effect of altering a variety of variables.)

Calculating the direction of the joint reaction force

Two methods may be used to calculate the joint reaction force. Either a graphical method or a mathematical method. For both methods it is necessary to know the inclination to the vertical of the abductor muscle vector, the abductor force (calculated by the principle of moments), the length of the abductor lever arm and the length of the body weight lever arm and body weight.

The abductors are divided into two groups: rectus femoris, sartorius and tensor fascia lata with a resultant force inclined at 5.5° to the vertical in a caudomedial direction and a second group comprising gluteus medius, minimus and piriformis with a resultant force inclined at 29.3° to the vertical in a caudolateral direction. The overall abductor resultant force was calculated to be inclined at 21° to the vertical in a caudolateral direction with a moment arm of 4 cm from fulcrum of the hip joint.

The body weight lever arm is calculated as 10.99 cm in Fischer's phase 16 of the gait cycle, that is monopedial stance on the right leg with the left leg half way through the swing phase. At this phase the centre gravity of body weight is in the same coronal plane as the centre of the head of the femur. With these data the magnitude and the direction of the joint reaction force can be calculated either by graphical or mathematical methods.

The theorem of the triangle of forces

The theorem of the triangle of forces states that if three forces acting on a body are in equilibrium then they can be represented in magnitude and in direction by the sides of a triangle taken in rotation, i.e. with the direction following around in the same sense.

Thus, to construct a triangle of forces for the lever system in Figure 5.8, a vertical line to the scale of force *W*, i.e. body weight is drawn. From the lower end of this line, a line to the scale of the abductor force is drawn at an angle of 21° (the inclination of the abductor muscle vector, *M*). A line drawn connecting these two lines represents the joint reaction force and its angle to the vertical (Fig. 5.9).

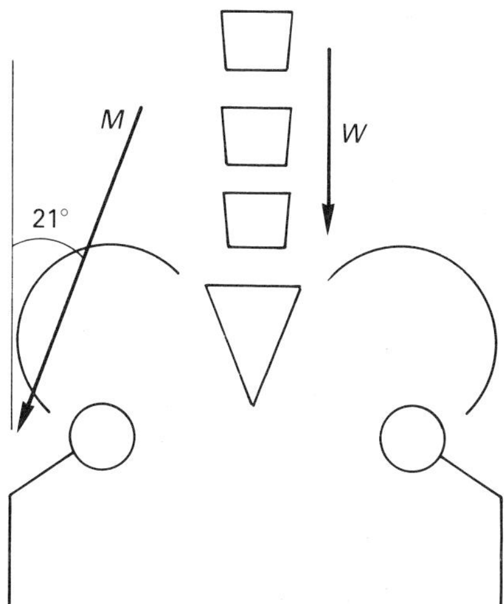

Figure 5.8 Lever system of forces about the hip. *W*, body weight; *M*, the abductor force which pulls at an angle of 21°

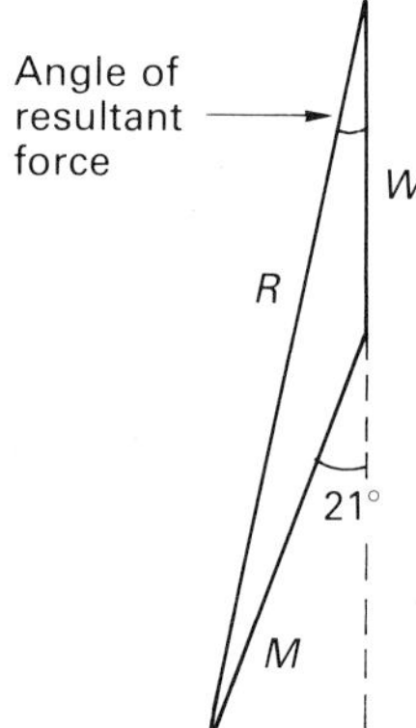

Figure 5.9 Resolution and calculation of the resultant force using triangulation. *W*, body weight; *M*, the abductor muscle vector at 21°. This allows the angle and magnitude of *R*, the resultant force, to be calculated.

The mechanical stresses on the normal hip joint

Before studying the effects of altering the configuration of the upper femur we must look at the mechanical stresses on the normal hip joint. The resultant force on the hip joint creates compressive forces. The effect on the hip joint depends upon:

1 The magnitude of the force.
2 The direction of the force.
3 Size of the weight-bearing area.

Bone and cartilage both respond to mechanical stressing. Excessive loading or stressing below a physiological threshold results in bone and cartilage loss (Fig. 5.10).

Subchondral sclerosis of the roof of the acetabulum (the sourcil) as seen in a plain X-ray of the hip implies that the joint stresses are evenly distributed (Fig. 5.11).

This, however, conflicts with the bell-shaped curve that is seen in an experimental model of a ball and socket joint where the stress distribution is maximal along the line of action of the resultant force and declines towards the periphery (Fig. 5.12).

To explain the sourcil normal articular cartilage must therefore be elastic and have the ability to distribute load applied across the hip joint. This is aided by the anatomy of the hip joint with the femoral head diameter being slightly larger than the acetabular socket. With minimal load, weight-bearing occurs at the periphery and with increasing load so the centre is loaded. In disease, as articular cartilage is lost then point loading occurs due to the failure of this stress distribution. This is reflected by the development of a bell-shaped area of subchondral sclerosis over the site of maximal loading which may occur anywhere in the acetabulum (Fig. 5.13).

In the normal hip the joint reaction force is inclined 16° to the vertical acting through the centre of the femoral head (Fig. 5.14a) and stresses are distributed

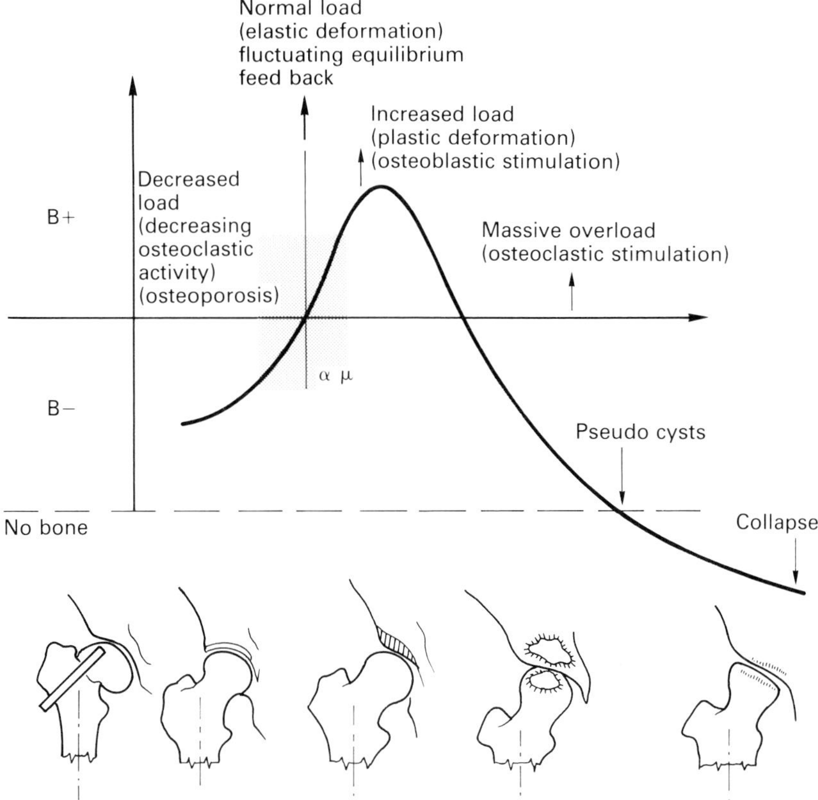

Figure 5.10 A relationship between bone activity and stress, (α). B− is bone reabsorption and B+ is bone formation. Where the hip is subject to normal load bone deposition is balanced by bone reabsorption at a point of minimum stress (α and μ). Where stress is reduced or increased the balance is upset and bone and cartilage loss results.

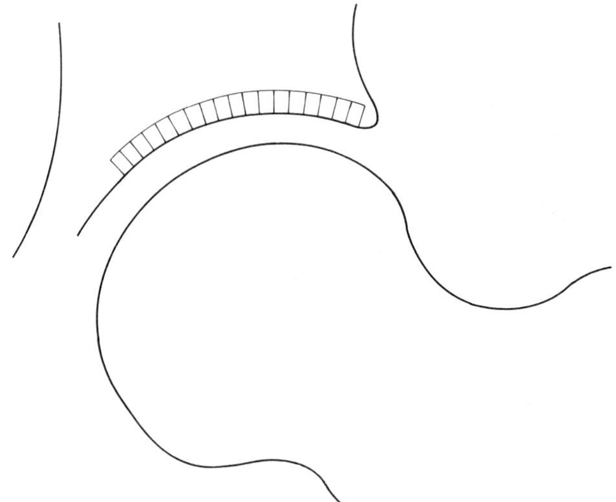

Figure 5.11 The sourcil in a normal hip shows normal joint stress distribution.

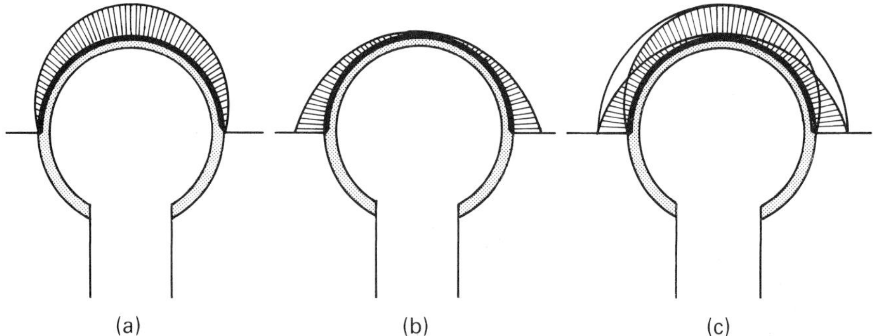

Figure 5.12 (a) Stress distribution in a normal ball and socket joint; (b) where the ball and socket joint is slightly incongruent stresses are greater at the periphery; (c) when the incongruent ball and socket joint is loaded the stresses are evenly distributed (Adapted from Kummer, 1978)

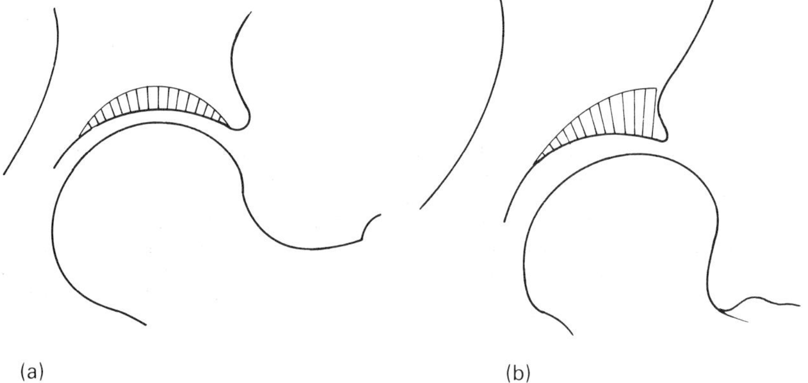

Figure 5.13 Diagram to show the movement of the bell-shaped area of some subchondral sclerosis reflects the stress distribution in the hip. (a) normal appearance, (b) abnormal, with peak loading laterally and resultant movement laterally of the area of sclerosis

evenly across the joint as reflected by the normal acetabular sourcil. In coxa vara, the abductor moment arm is increased. The resultant force is therefore both reduced and less vertical (Fig. 5.14b). The weight-bearing area is unchanged so the reduced forces remain evenly distributed.

In coxa valga, the moment arm of the abductor is reduced, consequently the abductor force is increased and more vertical. This results in the joint reaction force coming to lie closer to the vertical. So long as the joint force still passes through the centre of the joint, the stresses are distributed evenly albeit higher due to the reduced abductor moment arm and over a smaller weight bearing area due to uncovering of the lateral femoral head (Fig. 5.14c).

If the line of action of the resultant force is displaced and does not pass through the weight-bearing centre as occurs when the hip is subluxed, then the distribution of the stress becomes uneven as the load must be borne equally on each side of the line of action of the force. Thus the compressive forces are greater on the side nearest to the edge of the acetabulum due to the reduced weight bearing surface (Fig. 5.15). This leads to a dense triangle of subchondral sclerosis laterally. If the forces increase beyond the tolerance

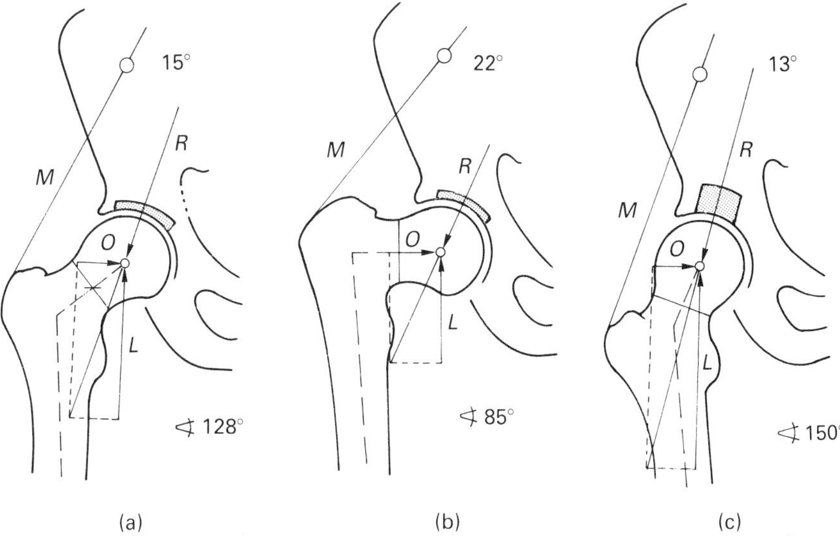

Figure 5.14 (*a*) Normal alignment of hip, *M*, abductor force; *R*, resultant force. The area of subchondral sclerosis indicates normal stress. (*b*) Coxa vara. The abductor moment arm is increased (*M*) and the resultant forces are therefore both reduced and less vertical. As the weight bearing area is maintained the reduced resultant force is evenly distributed. (*c*) Coxa valga. The moment arm of the abductors is reduced (*M*) and as a result the abductor force is increased and more vertical causing a more vertical and increased resultant force (*R*). The joint force is increased and as the head becomes uncovered laterally, there is a small weight bearing area which causes increased stress.

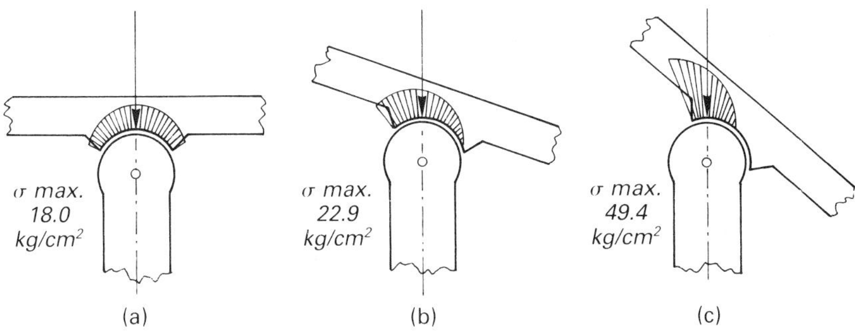

Figure 5.15 The resultant sourcil at (a) central loading and (b, c) loading off centre where α maximum equals maximum stress. (Adapted from Pauwels, 1961)

of bone then a subchondral cyst forms. This may be relieved by lateral new bone formation or the surgical provision of a lateral roof. Medial osteophyte formation, though increasing the overall contact area does not reduce the lateral compressive forces, indeed by causing further lateral migration of the head it increases the lateral compressive force (Fig. 5.16).

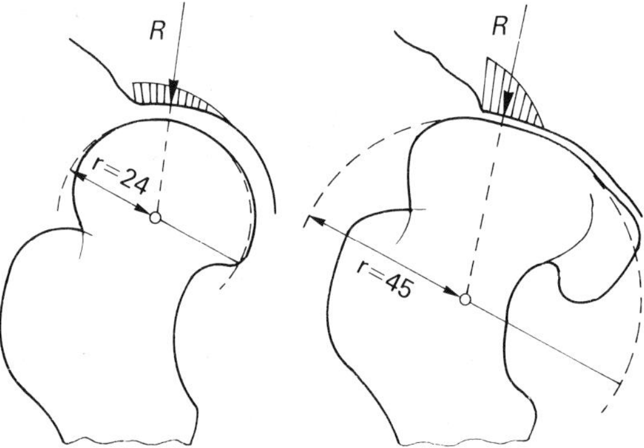

Figure 5.16 Effect of medial osteophyte formation which produces further lateral migration of the head increasing lateral compressive forces.

In the immature skeleton the normal capital epiphysis is perpendicular to the resultant force so an even distribution of the compressive force across the growth plate stimulates uniform growth. But, following varus osteotomy, the medial compressive forces are increased, this stimulates medial bone growth, consequently with time, the varus alignment of the neck grows out (Fig 5.17). In congenital coxa vara this fails to occur, as the physis is unable to respond to the asymmetrical loading and the coxa vara persists.

Biomechanical treatment of osteoarthritis

It is the ability for remodelling to occur in hypertrophic osteoarthritis that is the basis for its biomechanical treatment. Atrophic osteoarthritis does not respond well to alteration of the hip mechanics.

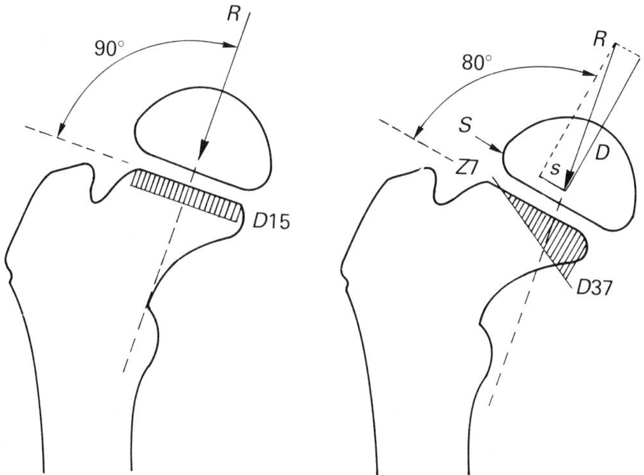

Figure 5.17 The effect of varus osteotomy causing a change in the resultant force *R*. When inclined to the growth plate the uneven distribution produces stresses on the plate encouraging medial bone growth. *S*, shearing force; *D*, compression, *Z*, tension.

Treatment may be achieved either by increasing the resistance of tissues to the compressive loads or by altering the anatomy to reduce effectively the compressive loads. Currently we have no methods available to increase tissue resistance to loading. Therapy must therefore be directed at reducing joint loads. This may be by reducing the compressive forces directly or by increasing the weight-bearing area, and thereby reducing the load per unit area or ideally by combination of the two.

Hanging hip procedure

Tenotomizing the muscles around the hip joint reduces their muscle power and therefore the joint reaction force is reduced (Fig. 5.18). The adductors and ilio-psoas are divided medially and the fascia lata, gluteus medius and minimus laterally. This procedure is advocated in patients with osteoarthritis or rheumatoid arthritis who have congruent joint surfaces without evidence of medial or lateral concentration of joint forces as seen by local dense subchondral sclerosis in the acetabulum.

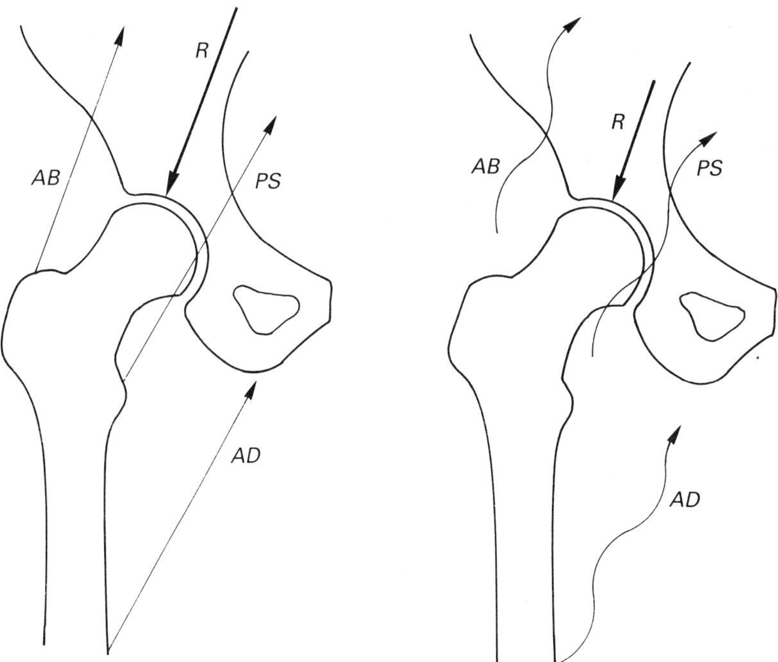

Figure 5.18 The hanging hip procedure reduces the resultant force R acting on the joint as a result of release of the abductors (AB), adductors (AD) and psoas.

Varus intertrochanteric osteotomy (adduction osteotomy)

A varus osteotomy lengthens the abductor lever arm so reducing the force required by the abductor muscles (Fig. 5.19). Moreover, it changes the direction of the abductor force to a less vertical force which, in turn, displaces medially the resultant force of the hip. The effect of this

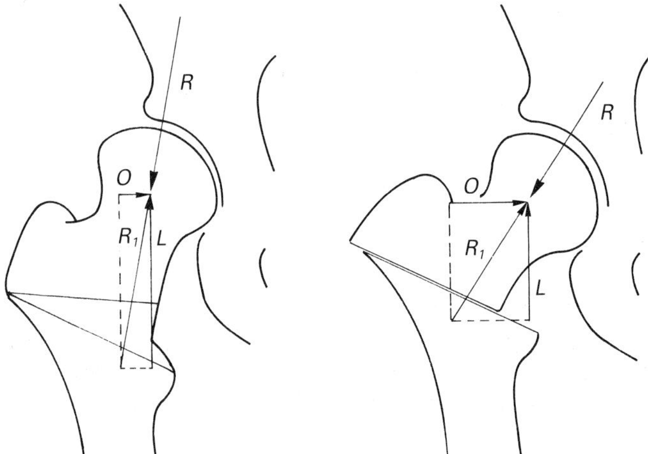

Figure 5.19 The resultant force R is decreased across the joint by varus intertrochanteric osteotomy as the articular weight bearing surface is increased and the abductor force, M, is reduced and made more horizontal.

is to increase the weight bearing area of the hip (even if the contact area is not increased) as the force must be distributed evenly each side of the resultant force.

A varus osteotomy therefore reduces the resultant force and increases the weight-bearing area of the hip. The altered joint mechanics following varus osteotomy is reflected by a change of the femoral trabecular pattern. In osteoarthritic patients the lateral triangle of subchondral sclerosis in the acetabulum is replaced by a more normal thin subchondral sclerosis, reflecting healing (Fig. 5.20).

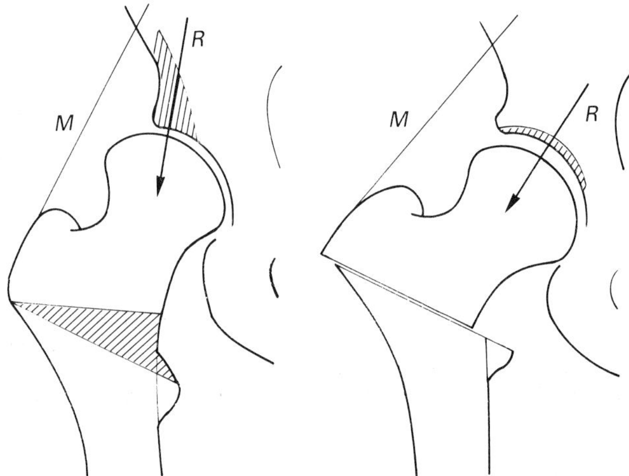

Figure 5.20 Action of varus osteotomy increasing transverse component Q and decreasing longitudinal component L. R_1, the reaction force; R, the resultant compressive force.

Indications include:

1 Symptomatic coxa valga with subluxation
2 Osteoarthritic patients with triangular subchondral sclerosis at the acetabulum margin.

An essential prerequisite is that the hip is congruent in full abduction, otherwise an increase in weight-bearing area will not occur and there will be point loading.

Effects of osteotomy on the knee

A varus osteotomy lateralizes the greater trochanter and the femoral shaft. Lateralizing the femoral shaft results in a varus knee (Fig. 5.21). The femoral shaft must be displaced medially to restore its original relationship with the pelvis.

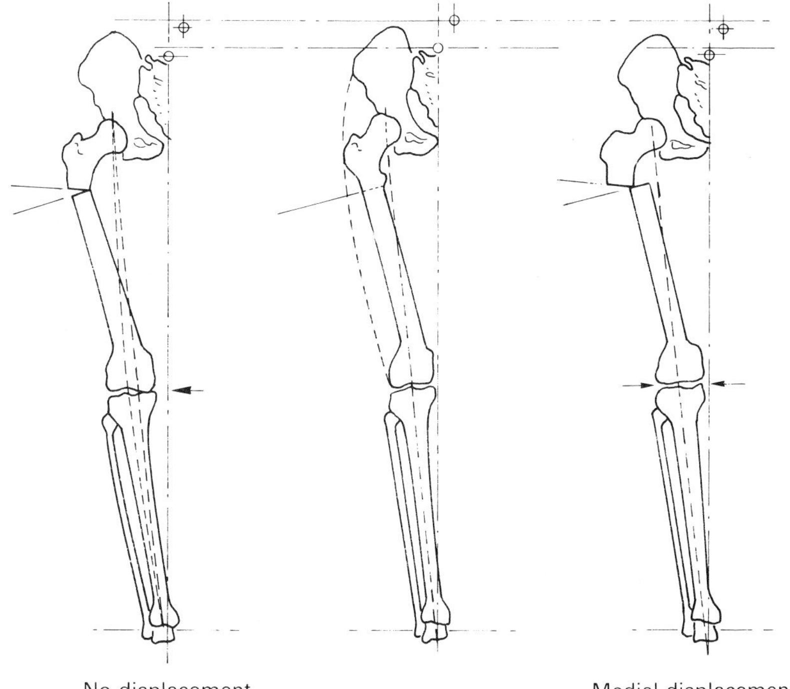

Figure 5.21 The effect of varus intertrochanteric osteotomy on the knee. The centre of rotation of the femoral head is displaced medially. As a result the mechanical axis of the limb is medialized and the medial femoral and tibial condyles are overloaded. When the medial shift of the femoral shaft is combined with varus osteotomy the mechanical axis is maintained through the centre of the knee.

Valgus intertrochanteric osteotomy (abduction osteotomy)

In a normal hip, a valgus osteotomy shortens the abductor lever arm and displaces the joint reaction force closer to the edge of the acetabulum, neither of which is desirable. But, in a grossly deformed head the medial osteophyte displaces the femoral head laterally. The joint reaction force continues to act through the centre of the head and is therefore displaced closer to the edge of the acetabulum. Consequently, though the contact area is greater, the weight-bearing surface is reduced. A valgus osteotomy swings the centre of the femoral head back medially and thereby increases the abductor lever arm and reduces the body weight lever arm. The lateral prominent head displaces the line of pull of the abductors and increases further the abductor lever arm. The joint reaction force is moved medially so the weight-bearing area is increased and the medial femoral osteophyte can be utilized as a weight-bearing structure (Fig. 5.22).

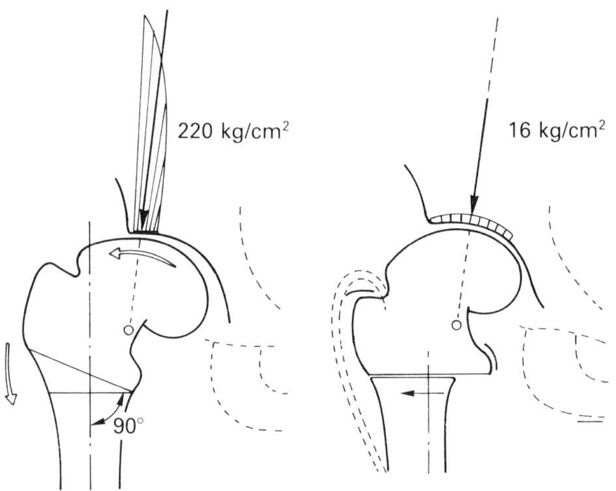

Figure 5.22 Valgus intertrochanteric osteotomy. The formation of the osteophyte over the medial aspect of the femoral head allows, with rotation, the joint surface to be increased by valgus realignment. When combined with tendotomy of the abductor, adductor and iliopsoas muscles, the load transmitted across the hip is reduced. This, coupled with the enlargement of the weight-bearing surface reduces the joint force across the hip.

Indications include:

1 A joint with improved joint congruency in adduction.
2 A joint that demonstrates hinge abduction.

This is most frequently seen as a feature of Perthes disease. Flattening of the femoral head and calcification lateral to the femoral head leads to impingement of the lateral margin of the head and acetabulum on abduction. The medial joint space is seen to open as the head hinges laterally (Fig. 5.23).

A valgus osteotomy aims to increase the weight bearing area by improving joint congruency.

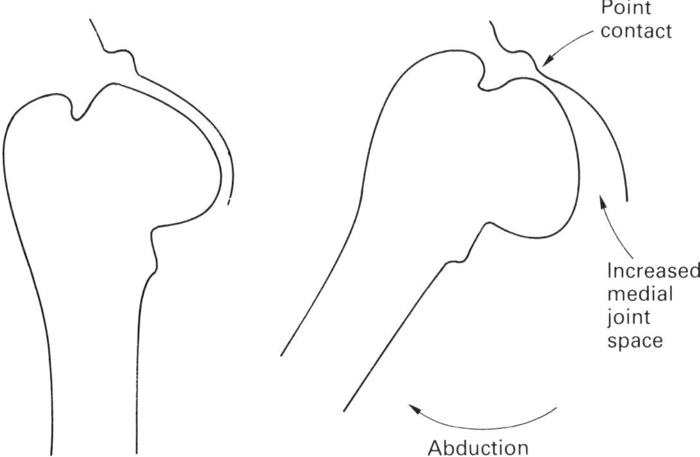

Figure 5.23 Hinge abduction where the lateral part of the femoral head impinges on the lateral margin and acetabulum to produce opening of the medial joint space.

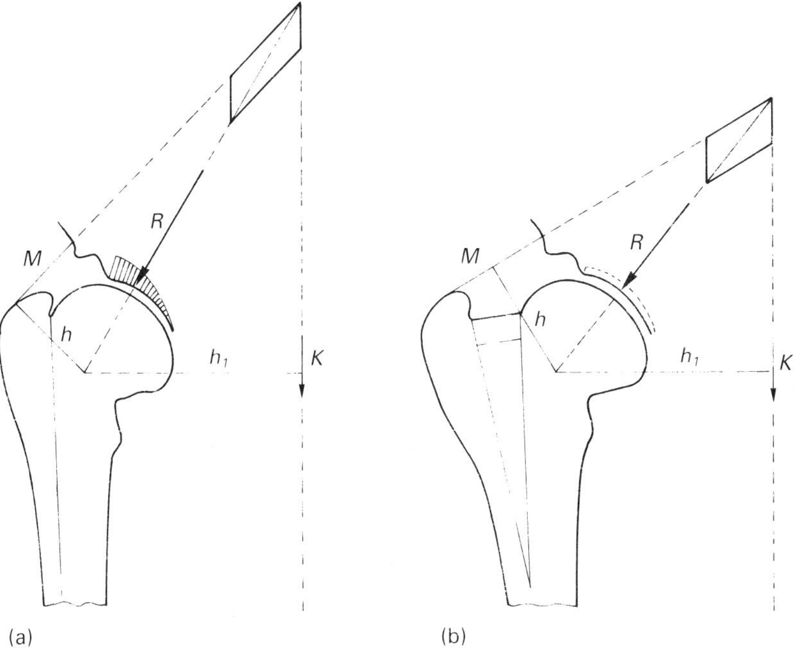

Figure 5.24 Lateral displacement of the greater trochanter. This has the effect of reducing the abductor force (M) and enlarges the weight bearing surface of the joint. K, weight of body; h_1, the lever arm of force K; h, the lever arm of force M; R, the resultant of forces K and M. Effects seen before (a) and after operation (b).

Lateral displacement of the greater trochanter

Lateral displacement of the greater trochanter increases the abductor lever arm, reducing the abductor force. This displaces the joint reaction force medially, so increasing the weight bearing surface (Fig. 24).

Indications include:

1 Osteoarthritis with joint congruency in all positions where a varus osteotomy will result in the greater trochanter being higher than the femoral head.
2 Patients with severe coxa valga with a short abductor lever arm.

Distal displacement of the greater trochanter

This tightens the abductors leading to an increase in joint pressure. In addition it shortens the abductor lever arm. So the resultant force is displaced closer to the edge of the acetabulum which potentially reduces the weight-bearing surface (Fig. 5.25).

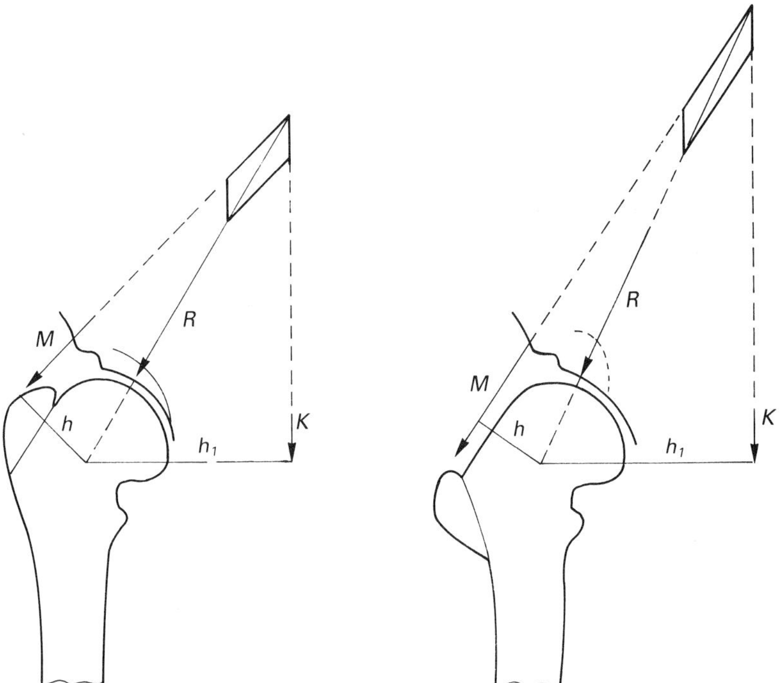

Figure 5.25 Distal displacement of the greater trochanter. This may increase the force R as a result of increasing the abductor force M and also reducing the weight bearing joint surface. K, the weight of the body; h_1, the lever arm of K, M the force exerted by the abductor muscles; h the lever arm of M; R, a resultant of K and M.

Aspects of biomechanics related to total hip replacement

Femoral component design

Femoral component design determines how the joint reaction force is transferred to the proximal femur. This is influenced by:

1 The geometry of the prosthesis.
2 The material property of the prosthesis.

Geometry of the femoral component

The following variables will be considered:

1 Stem size
2 Stem length
3 Cross-sectional shape
4 Collars

1. Stem size (Fig. 5.26).

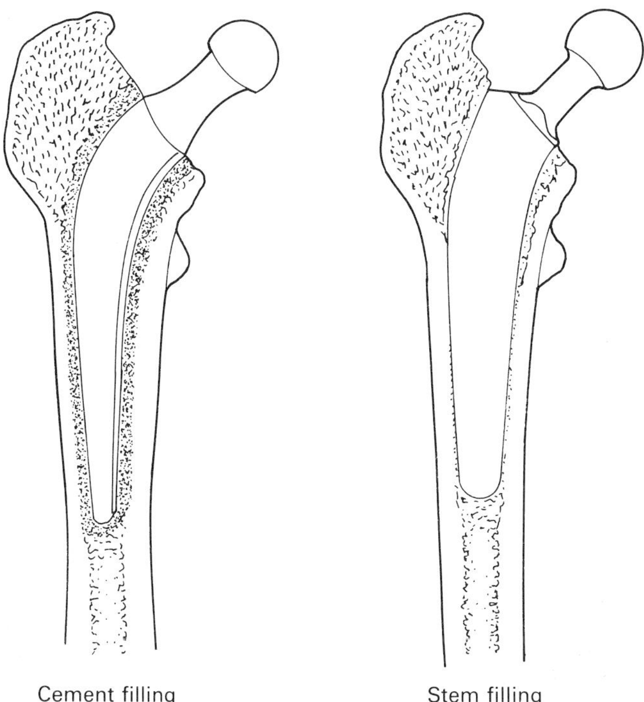

Figure 5.26 A traditionally sized femoral component requires a larger cement mantle than a proportionally larger implant, fitting a higher proportion of the medullary canal

1 A thin stem requires a cement mantle to transfer load.
2 A large stem requires cement only to act as a grout.

Thin stems are therefore highly dependent upon the structural properties of the cement-to-load transfer. Cementing technique is therefore extremely important.

In contrast, the larger femoral components rely upon the wedge fit of the prosthesis into the proximal femur producing multiple contact points of the stem on the cortical bone. Consequently the cement is thin and there are multiple areas where the stem will make direct contact onto the endosteal surface of the femur. Cementing technique is therefore less critical.

Cement stresses are greater the thinner the cement mantle is. (Fig. 5.27). Incomplete cement mantles are susceptible to mechanical failure.

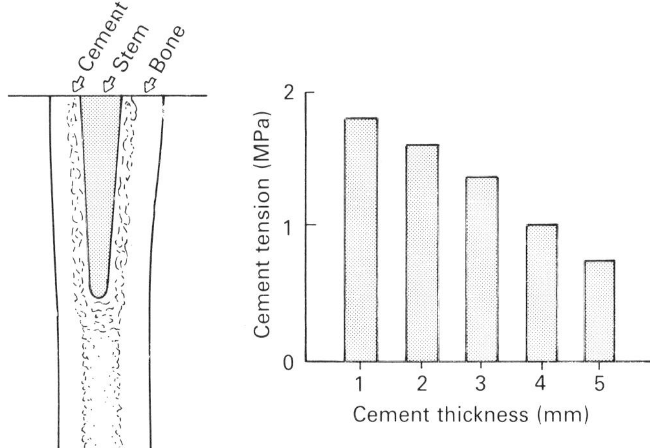

Figure 5.27 As the cement thickness increases the tension within the cement decreases (Adapted from Kwak B.M., Lim, O.K., Kim, Y.Y. et al, *Int. Orthop,* 2: 315, 1979)

2. Stem length

The longer the stem the greater the distribution of load and therefore stresses per unit area applied to the cement mantle are reduced. A stem should engage the femoral isthmus to increase the stability of the prosthesis and also to direct its alignment.

3. Cross sectional area (fig. 3).

1 Stem cross-sections with low bending stiffness and sharp medial aspects lead to high stresses on the cement mantle and consequently are prone to early failure.
2 High bending stiffness and smooth contours and shapes with good compressive rather than tensile loadings are advantageous.
3 Implants have to resist torsional loads (Fig. 5.29). Non-circular cross-sectional designs resist torsional loads better. Much of the torsional load

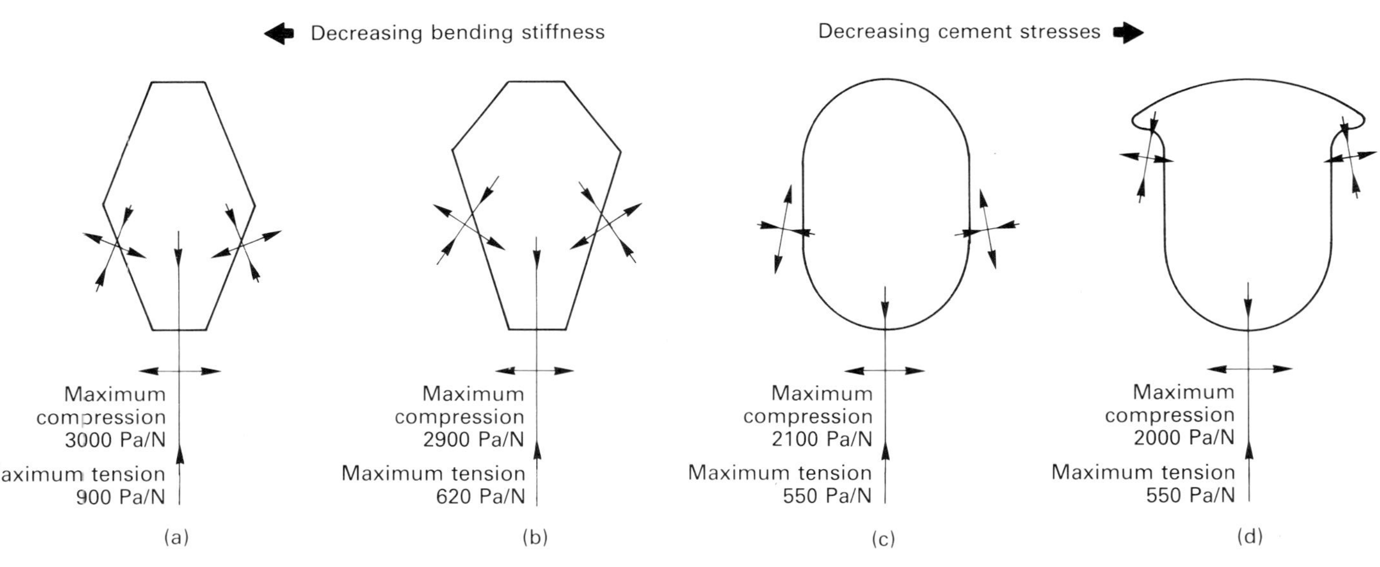

Figure 5.28 Different cross-sectional shapes (a)–(d) produce varying tensile and compressive stresses on the proximal bone cement.

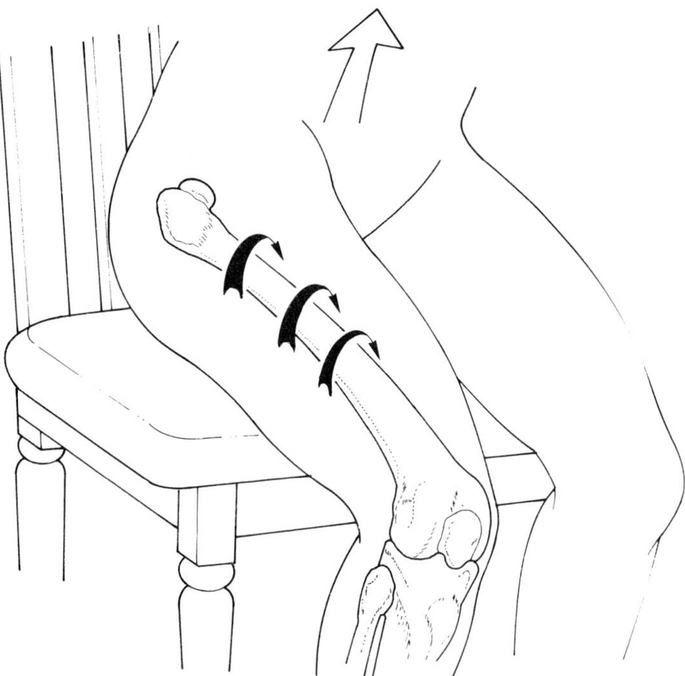

Figure 5.29 Peak torsional forces often act on the proximal femur when rising from a chair

is resisted by the proximal stem design, but as the proximal femur is relatively flexible (a thin cortical shell) distal stem design needs to supplement this. It is also necessary to taper the distal stem to disperse stresses to avoid any peak stress raisers at the tip of the prosthesis.

4. *Collars* (Fig. 5.30)

1 Proponents of collared designs claim that the prosthetic collar transfers load onto the medial femoral cortex proximally. In addition, they postulate that even if the collar does not transfer load, in cemented designs it helps to improve cement pressurization at the time of implantation.
2 Collarless designs: Proponents of this design claim that collars stop sinkage. This mitigates against a tight interlock re-forming if the prosthesis becomes loose. The collar instead of allowing sinkage of the prosthesis acts as a calcar pivot and the prosthesis fails as shown by the method in Figure 5.31.

Recent finite element analysis has suggested that even if collars do increase load transfer to the medial cortex this only occurs with the more flexible prosthetic designs, particularly those of titanium. In press-fit designs, which attempt to fill the proximal medulla, little further loading will occur by the addition of a collar. It is clearly crucial if load transfer is to occur through the collar that there is direct bony contact. This is highly dependent upon operator technique.

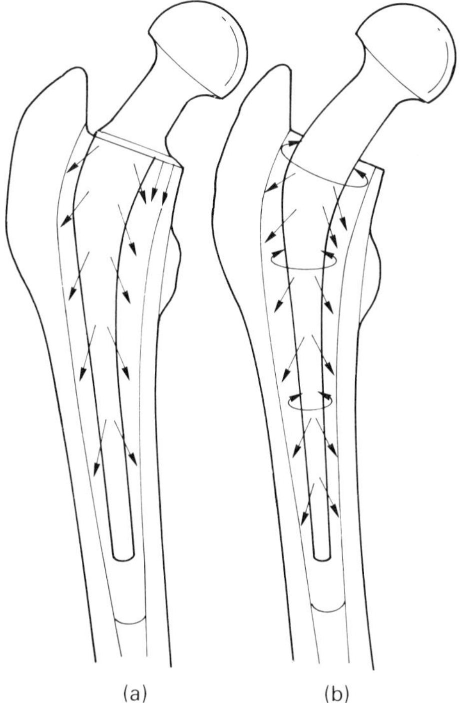

Figure 5.30 (a) The addition of a collar allows loading through the calcar. (b) In the absence of a collar there are increased hoop stresses produced

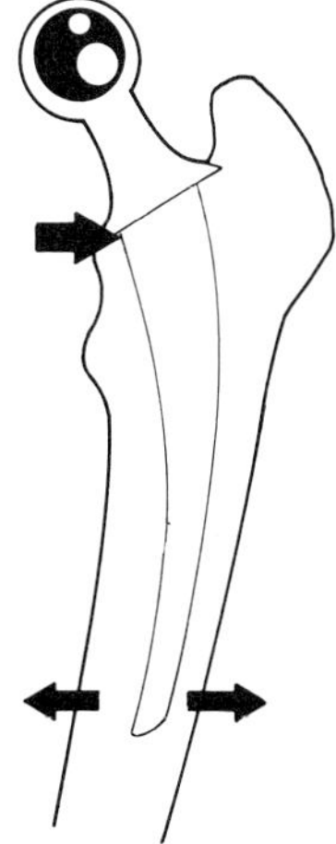

Figure 5.31 The addition of a collar can produce a 'windscreen wiper effect' at the distal prosthesis tip

Calcar Loading

In the normal hip the calcar is loaded. All prostheses reduce calcar loading. With collared prostheses this reduction is to 50–70% of the normal physiological load. In collarless designs it is down to approximately 10–20% of the physiological load. The role of calcar reabsorption as an aetiology of femoral loosening is unclear but anything to increase proximal cement load and thereby to increase calcar loading is probably advantageous. Collars should not prevent access to the anterior or posterior aspects of the shaft of the femur in case removal of the implant is necessary.

Material properties

1. Modules of elasticity (Fig. 5.32)

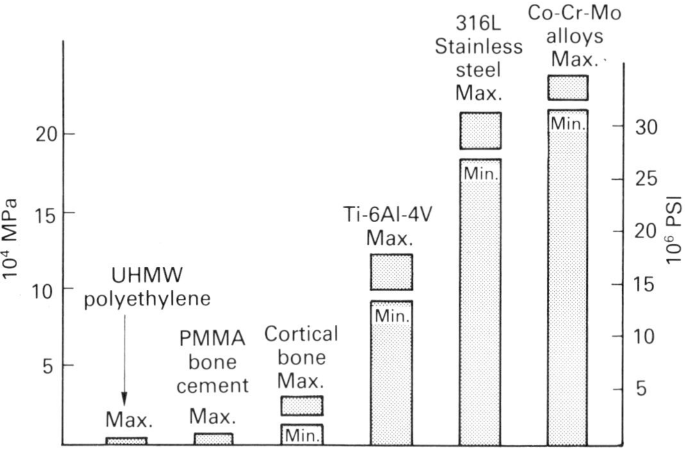

Figure 5.32 The modulus of elasticity varies considerably between the stainless steel and alloy and implantation materials and that of cortical bone

The stiffness of any femoral component design depends upon the modulus of elasticity of the constituent material, the cross sectional area of stem and the stem shape. The more flexible the prosthesis the greater the proximal load transfer.

Titanium alloy has a low modulus of elasticity and, with its relative strength, permits designs of smaller cross-sectional area. It tends to transfer more load proximally. All metals used in implant manufacture have a higher modulus of elasticity than cortical bone (titanium four- to six-fold, CoCr alloys 10-fold). Proponents of proximal load transfer favour more flexible implants. Those who are worried about proximal overload and bone reabsorption would veer towards more rigid materials.

2. Material Strength (Fig. 5.33)

Fatigue failure is the most relevant to implant design and is defined as a progressive fracture of a part subjected to repeated cyclical or fluctuating

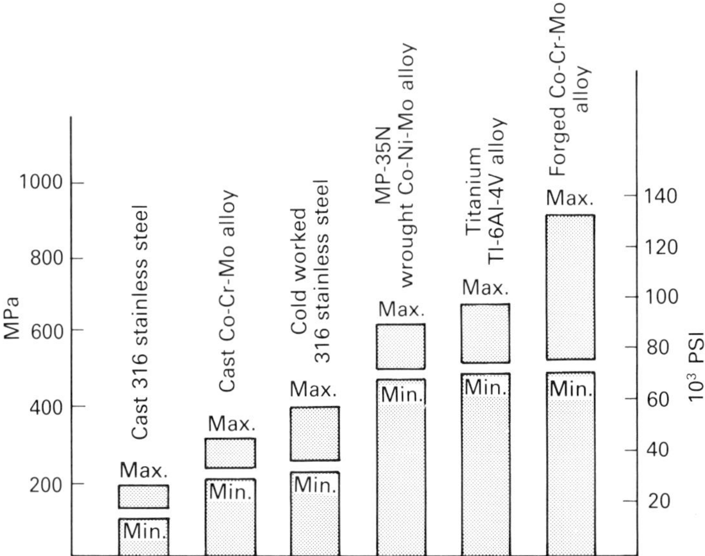

Figure 5.33 Graph of material fatigue strength for various orthopaedic biomaterials (per 10 million cycles)

loads. The fatigue strength of an implant material is altered permanently by heating. Sintering, a method of applying a porous coat by metallurgic bonding which involves heating to high temperatures, considerably reduces the fatigue strength of the implant (Fig. 5.34). Alternative techniques that do not require heating the prosthesis to high temperatures would be advantageous.

Femoral head size

Historically head diameters have ranged from 22 to 32 mm. The 22 mm head size was advocated by Sir John Charnley (1979) as it produces less frictional torque and hence lower stresses on the acetabular/cement/bone interface, the so called LFA (low friction arthroplasty) principle. Small femoral heads however have high contact stresses resulting in wear of the acetabular polyethylene insert. Small femoral heads are also more liable to subluxation/dislocation. This is a consequence of the reduced range of movement inherent in the design due to impingement of the neck against the acetabulum earlier than would occur with a larger head (Fig. 5.35).

At the other extreme a 32 mm head generates greater frictional forces and consequently generates greater stresses on the acetabulum/cement/bone interfaces. The increased weight-bearing surface of the polyethylene results in greater wear debris than in 22 mm heads in spite of the lower contact stresses. The larger head has an increased range of movement and is thus a more stable implant.

Currently surgeons favour a middle course selecting a 26 mm or 28 mm head size in an attempt to gain the advantages of both.

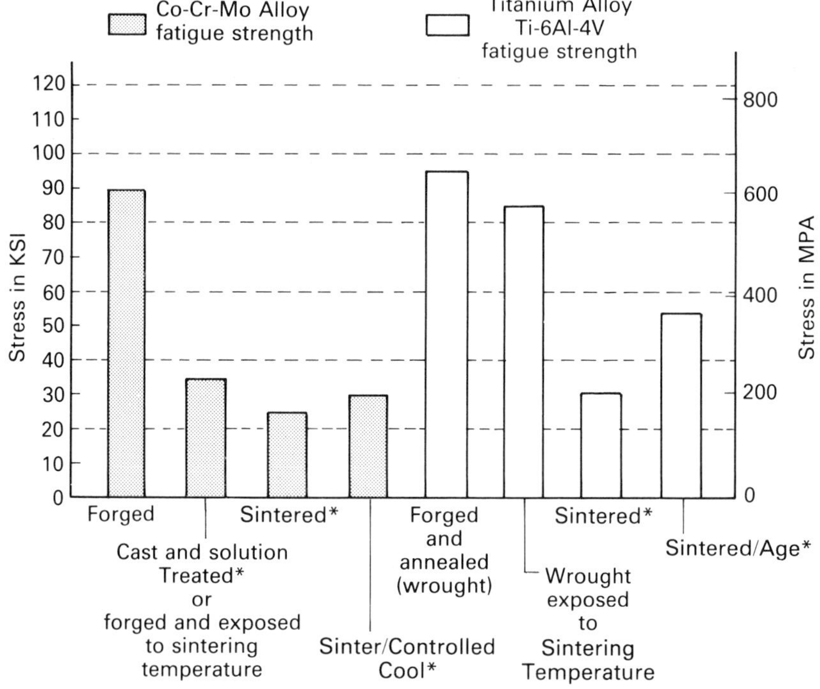

Figure 5.34 Fatigue strength per 10 million cycles of cobalt chrome alloy and titanium alloy in forged and sintered conditions (Reproduced by permission of *Clin. Orthop*, 176: 42–51, 1983)

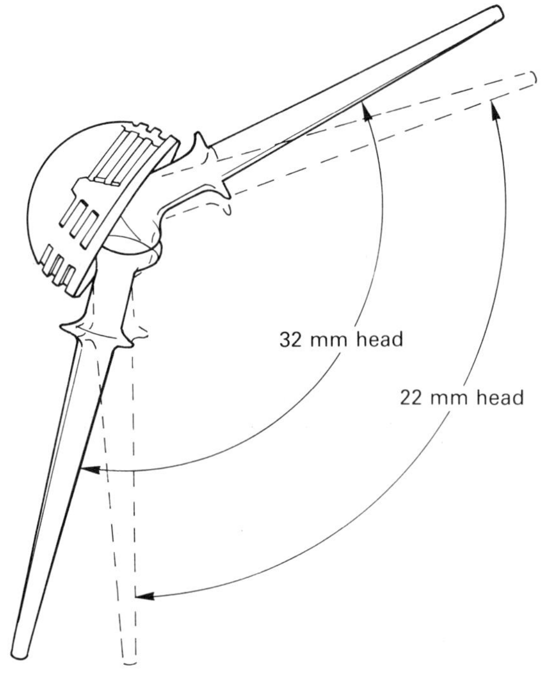

Figure 5.35 Impingement occurs earlier with a 22 mm head in comparison to a 32 mm head

Acetabular component design

Stress distribution to the acetabular region depends upon acetabular component design. High loading of the cement, subchondral bone and trabeculae may occur in the supramedial segment of the acetabulum (zone 1), whereas tensile stresses are produced in zone 3 (Fig. 5.36).

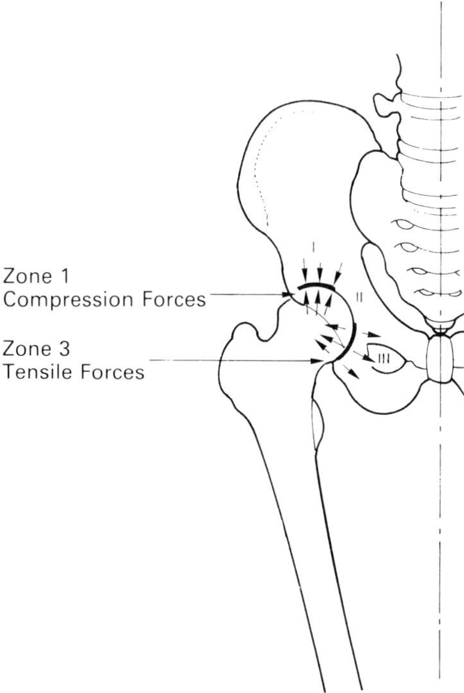

Figure 5.36 Tensile and compressive stresses are present in the acetabular component

Stress is highest in thin wall polyethylene components and is reduced in thicker walled prostheses.

Metal backing

Metal backing of the acetabular component stiffens the polyethylene implant, this results in a more even distribution in joint forces both to the cement mantle and to the subchondral bone.

Some peroperative biomechanical considerations

Femoral side

1 Selection of prosthetic offset
2 Valgus varus alignment

1. Selection of prosthetic offset

The relationship between the direction of the joint reaction force and the shaft of the femur constantly changes during ambulation. When the pelvis is elevated the direction of the reaction force becomes more vertical in relationship to the shaft of the femur (Fig. 5.37).

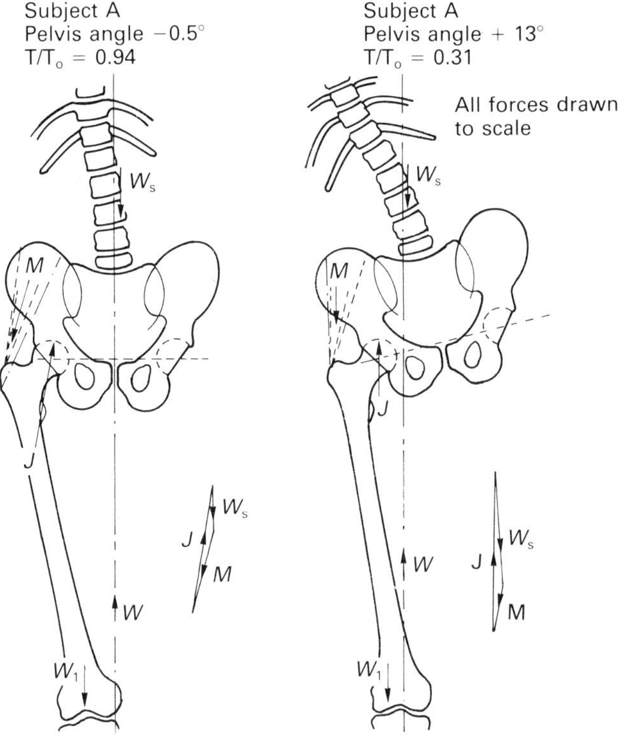

Figure 5.37 The alteration of the reaction forces as the pelvis shifts in normal ambulation. Note vector *J* the resultant force

When the pelvis drops, the angle between the joint reaction force and the femur increases (Fig. 5.38). By measuring the trabecular pattern of the upper shaft and neck of the femur it can be seen that the average angle of the joint reaction force is 20°. Following total hip replacement in order to reproduce physiological loading an average femoral offset of 45 mm would be required from the shaft of the femur (Fig. 5.39).

Increasing the offset increases the bending stresses on the implant and the stresses onto the prosthetic cement/bone interfaces. A decrease in the offset reduces the bending stresses on the implant and on the interfaces. However, in addition, it reduces the moment arm of the abductors which not only increases the joint reaction force but also displaces the joint reaction force more vertically thereby increasing the bending stresses. This can be negated by displacing the greater trochanter laterally (Fig. 5.40). Indeed Charnley's

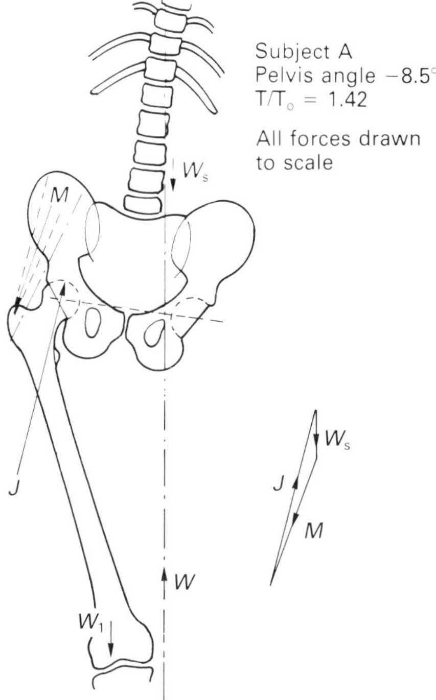

Figure 5.38 As the Trendelenburg position is adopted the joint reaction force increases (note the larger vector *J*)

original description of a total hip replacement was a hip reconstruction about an implant. In practice a 40° offset prosthesis is usually selected.

2. Valgus/varus alignment

The stem of an implant should be inserted into the neutral axis of the shaft of the femur. Varus alignment effectively increases the offset of the implant and increases the bending stresses on the prosthesis, risking implant failure. It also increases the stresses applied to the cement mantle leading to early femoral loosening.

Valgus alignment should also be avoided as it results in gross thinning of the medial cement mantle. The valgus position reduces the effective offset of the prosthesis and it would be better to select a prosthesis with a smaller offset which would be stronger and stiffer.

Lever arm of body weight

Medialization of the acetabulum reduces the lever arm of the body weight so reducing joint force, but if medialization of the acetabulum also involves medializing the insertion of the greater trochanter, the abductor force is more vertical so the joint reaction force becomes more vertical. The gains of reducing the moment arm of the body weight are far outweighed by the effect on the abductor force which is many times body weight. In addition, medialization also increases the bending moment on the prosthesis.

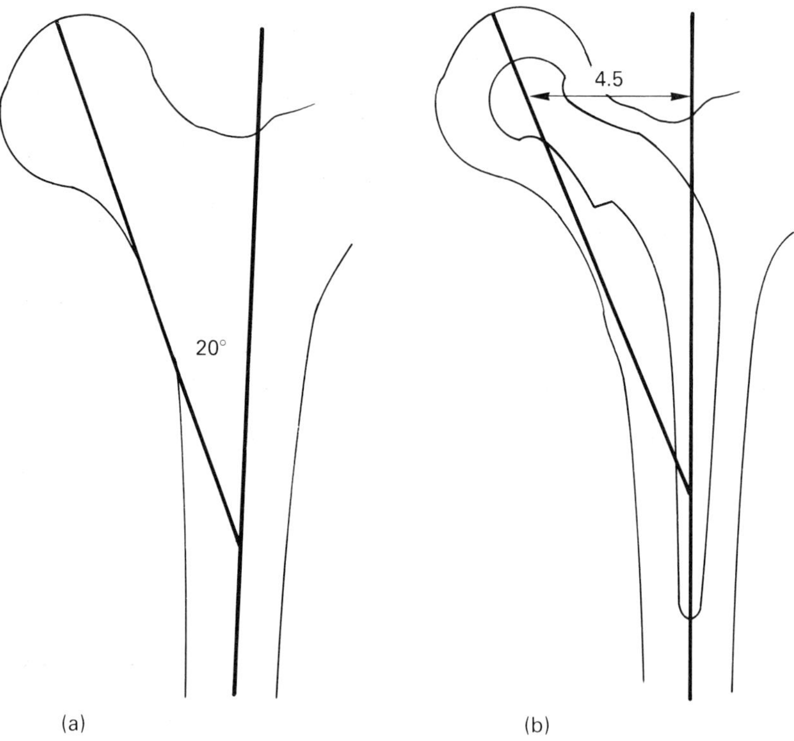

Figure 5.39 (a) The average resultant of the joint force is 20°. (b) This alignment is reproduced with an offset of 45 mm

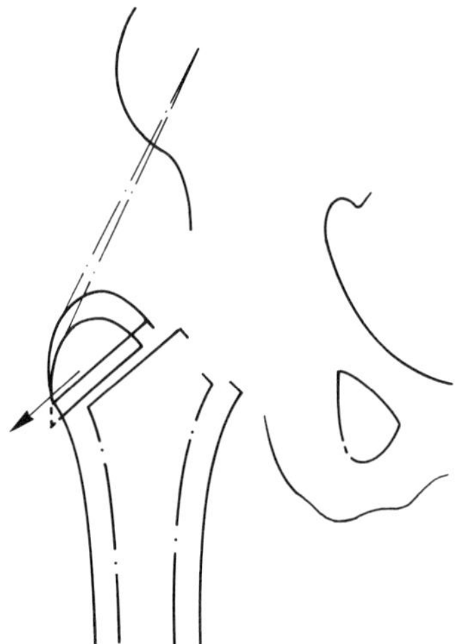

Figure 5.40 Bending stresses are reduced by displacing the greater trochanter laterally

Acetabular side

A more important consideration during surgery is to ensure full bony coverage of the acetabular component. The strongest available bone is the subchondral plate and it is onto this that the prosthesis should be seated rather than the weaker underlying cancellous bone. The aim of surgery should be to ream the minimal amount of subchondral bone to ensure full or at least 90% coverage of the acetabular component. Coverage is more important than the degree of abduction.

After total hip replacement the resultant force passes through the centre of the head of the femoral prosthesis. The acetabular component should be placed into the reamed position so that this resultant force bisects the subchondral plate, too medial or too lateral placement must be avoided.

If the acetabulum component is placed too laterally, excessive loading of the superolateral roof occurs and the prosthesis will fail due to progressive upper and lateral migration plus rotation. If the acetabulum is excessively medialized then the weaker medial wall of the acetabulum is unduly loaded. In order to achieve correct position of the acetabular component bone grafting or build up with cement may be required.

References

Bombelli, R. (1983) *Osteoarthritis of the Hip*. Springer-Verlag, Berlin, Heidelberg, New York.

Charnley, J. (1979) *Low Friction Arthroplasty of the Hip*. Springer-Verlag, Berlin, Heidelberg, New York.

Denham, R.A. (1959) Hip mechanics. *Journal of Bone and Joint Surgery*, **41B**, 550–557.

Podd, P. and Maquet, G.J. (1985) *Biomechanics of the Hip*. Springer-Verlag, Berlin, Heidelberg, New York and Tokyo.

Further reading

Nordin, M. (1989) *Basic Biomechanics of the Muscular-skeletal System*. Lea & Febiger, Philadelphia.

Stillwell, W.T. (1987) *The Art of Total Hip Arthroplasty*. Grune & Stratton Inc., Orlando, New York, San Diego and London.

Chapter 6

Biomechanics of the knee

Ian Dingwall

Definition of biomechanics

'The mechanical stresses to which living tissues are subjected under physiological conditions' (adapted from Maquet, 1976).

Force measurement

Measurement of forces about the knee joint is not yet possible by direct means. A number of indirect techniques including anatomical models, photoelastic models, mathematical analysis or clinico/radiological methods have been used experimentally and clinically to establish qualitative and quantitative force values about joint surfaces and their soft tissue restraints (Maquet, 1976).

Force resolution

By convention, forces are considered in two perpendicular planes about the knee joint. Forces are resolved in the plane of primary motion of the knee, that being the plane of flexion and extension. The perpendicular plane is the coronal plane and in that plane the effect of the single-legged stance on the forces developed are most easily appreciated. This is of relevance with respect to gait and understanding the mechanics of gait disturbance. Forces acting about the knee joint are therefore considered in the coronal plane to demonstrate the differences between the static standing two-legged posture and the single-legged weight-bearing stance which more closely approximates the stance phase of the gait cycle.

In the sagittal plane of the knee joint the effect of position of flexion, the position of the trunk and deformity, in the same plane, can be analysed.

Forces in the coronal plane

Bicondylar models of the knee have been produced showing a stress (contact force) distribution within the joint, which are modifications of the single condylar pattern of central concentration and diminishing stress to the periphery (Fig 6.1a). In the bicondylar model, the stress pattern is in two parts with the concentration for each condylar complex about the centre of the condyle and diminishing to the periphery and into the intercondylar area. The resultant force, however, runs in the intercondylar region between the centres of curvature of the condyles (Fig 6.1b).

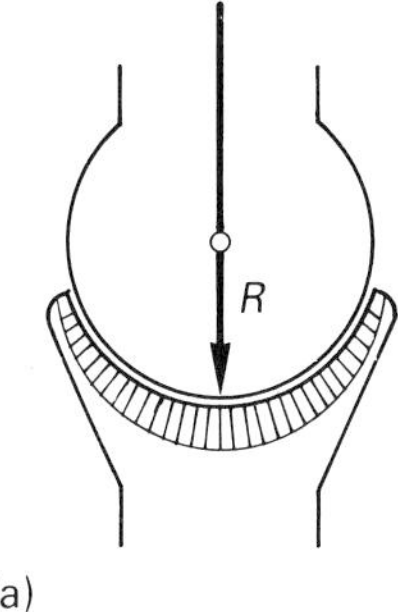

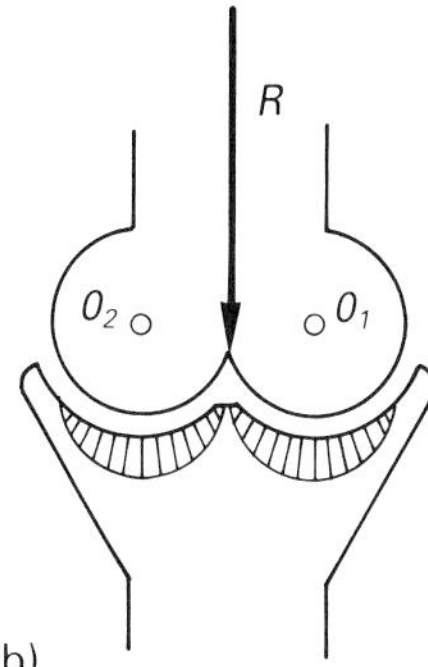

Figure 6.1 Single and bicondylar stress distribution pattern. R, resultant force, O_1, and O_2 are centres of the condyles. (From Maquet, 1976 by kind permission of Springer-Verlag, New York.)

Two-legged stance

In this position, the weight supported is 85% of total body weight, i.e. total body weight minus the weight of the legs below the knees. Therefore, the distribution of forces is equal between both knees with the value of 43% of body weight per knee.

Single-legged stance

In the single-legged stance, the tibio-femoral joint carries the whole body weight minus the mass of the leg below the knee on the supporting side. The partial body weight acting at a distance equal to that of the centre of gravity from the knee joint, must be balanced by a restraining lateral ligamentous/

muscular stay force to maintain the equilibrium (Fig. 6.2). To be in equilibrium, the resultant of these forces acting about the knee must pass between the centres of curvature of the two condyles, which approximates to the intercondylar notch of the femur. The magnitude and direction of the resultant force can be calculated if one knows the parameters – position of the centre of gravity, the partial body mass and the angles and distances of the lever arm (Fig. 6.2)

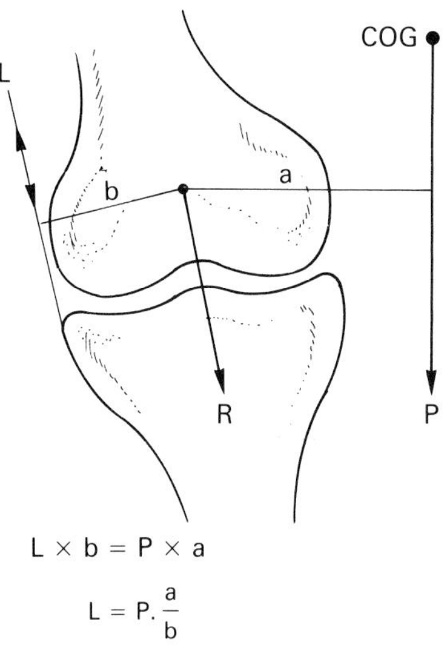

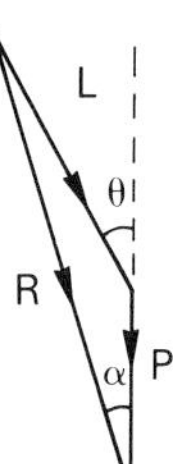

Magnitude and direction of resultant force

$$\text{Magnitude } R = \sqrt{P^2 + L^2 + 2\,P.L \cos\theta}$$

(θ = measured angle between L and P)

$$L \sin\theta = R \sin\alpha$$

$$\therefore \quad \sin\alpha = \frac{L}{R}\sin\theta$$

to give angle α of resultant (R) to the vertical.

Figure 6.2 Forces in the coronal plane. COG, centre of gravity; R, resultant force; P, body mass minus the weight of the supporting leg below the knee; L, lateral musculo-ligamentous restraining force

For the normal knee, in general terms the magnitude of the resultant force (R) is slightly more than twice the body weight. The angle of the resultant force to the vertical (α) is approximately 5° directed medially and distally.

The coronal alignment of the knee

A normal tibio-femoral angle (anatomical axis) is not precisely defined and estimates of the so-called normal angle differ, although the value varies from subject to subject. The average angle has been described as 7° valgus ± 2° (Denham and Bishop, 1978; Johnson, Leitl and Waugh, 1980), on the basis of studies of long leg, weight-bearing radiographs.

In the normal knee, the tibial plateau is at a 3° varus angle to the shaft of the tibia. The physiological valgus occurs as a result of the angulation of the femoral articular surfaces (condylar complexes) to the femoral shaft. One of the aims of surgical reconstruction is to restore the resultant force to the central intercondylar position, thereby restoring central balanced loading.

A line connecting the head of the femur and the centre of the dome of the talus, approximates the resultant force in the coronal plane. This line is known as the mechanical axis of the leg or the line of leg alignment (Fig. 6.3). In the normal knee, the line passes through the central intercondylar area of the tibial plateau, usually through the medial tibial spine.

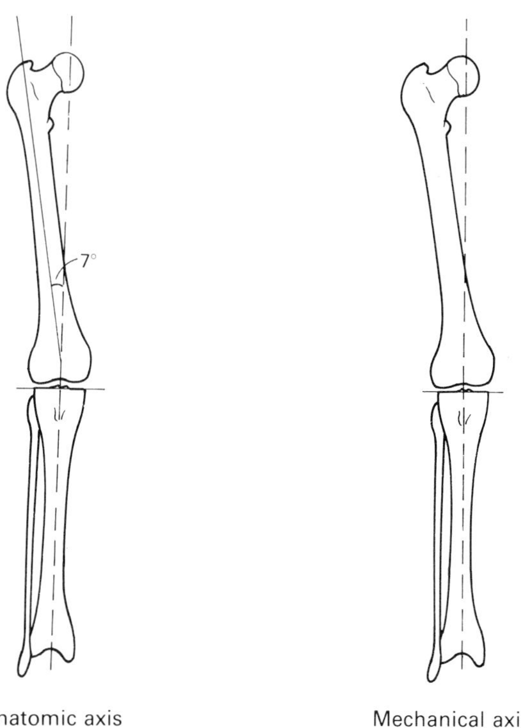

Figure 6.3 Reference axes of knee in the coronal plane

Whatever the anatomical axis measures for a given knee, the aim of reconstructive surgery is to restore the normal position of the mechanical axis.

The current generation of total knee replacements use intramedullary femoral jigs which are based on achieving a fixed anatomical axis, usually 7° of valgus. Achieving leg realignment, using the mechanical axis, under surgical conditions, is inaccurate. With the use of preoperative standing, long leg films, the angle of correction and hence the anatomical axis and jigs which are appropriate to restore the mechanical axis, can be assessed. Similarly, long leg films to assess alignment with reference to prosthesis survival are essential.

It has been shown that changes of tibio-femoral angle of as little as 10° (i.e. 3–4° varus, 17° valgus) can place the mechanical axis beyond the articular surface of the knee joint (Fig. 6.4). Beyond that point, the equilibrium of the joint is lost, whereby muscle action may be unable to close the gap in the non-weight bearing compartment and excessive tensions are developed in the

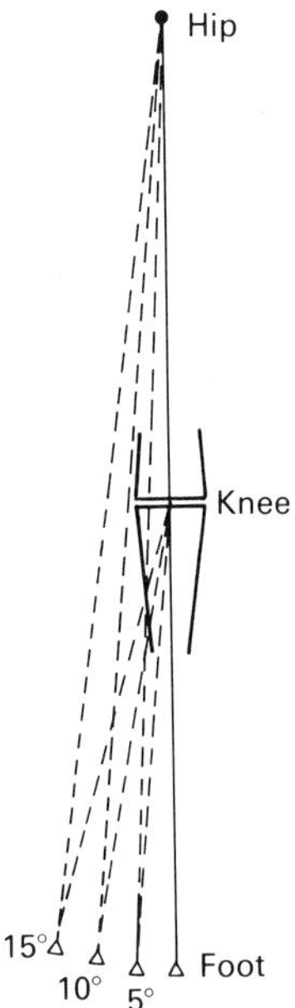

Figure 6.4 Small variations in the tibio-femoral angle produce large changes in the position of the line of leg alignment. (From Laskin, R., Denham, R. and Apley, A.G. (1984) *Knee Joint Replacement*, by kind permission of Springer-Verlag, New York.)

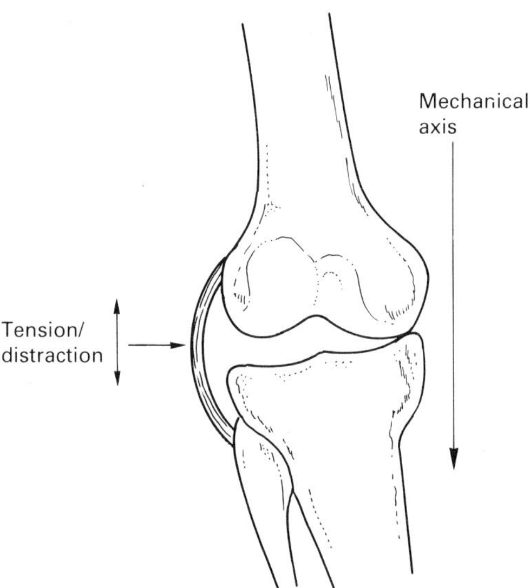

Figure 6.5 Loss of equilibrium

ligaments on the convex side (Fig. 6.5). In this situation, clinical deformity will become evident and symptoms may also be referable to the ligamentous structures on the convex aspect of the joint.

Clinically, it has been confirmed in static analysis studies that a varus deformity of 4° or a valgus deformity of 15° (as measured by anatomical axis) were the values of positions at which the entire compressive load would be carried by a single condylar complex, according to the relevant deformity (Johnson, Leitl and Waugh, 1980).

Limp

In a manner not dissimilar to the Trendelenburg limp of the hip, the patient with the varus knee deformity alters the position of the centre of gravity by moving the body mass towards the affected knee during stance phase. This reduces the effective lever arm of the body mass about the knee, in relation to the predicted value for the varus deformity. The balancing force (L) is reduced. The resultant force is reduced and equilibrium is restored (Fig. 6.6).

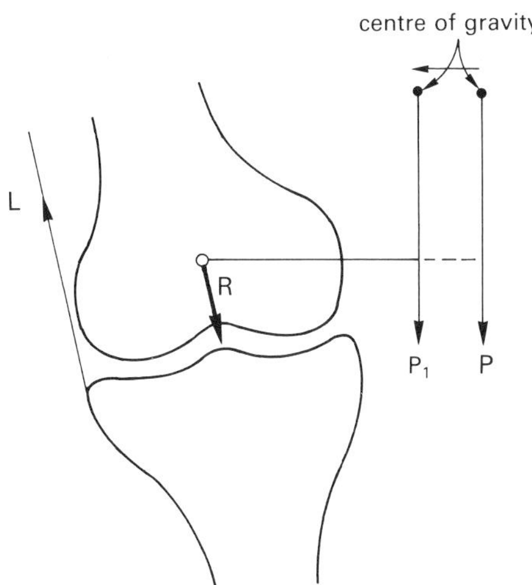

Figure 6.6 The effect of limp is to shift the centre of gravity thereby reducing the effective lever arm about the knee. P, old lever arm; P_1, new lever arm; R, resultant force; L, lateral musculo-ligamentous stay force

Use of walking stick

In a similar manner, by applying a force through a walking stick and the upper limb, on the opposite side from a varus knee, the effective lever arm length of the centre of gravity is shortened towards the affected knee. The walking stick also reduces displacement, consequent upon the deformity away from the affected knee. Both of these factors reduce the lateral stay and resultant forces. Limp is eliminated. Furthermore, the stick broadens the support base and improves balance and ensures a biomechanical equilibrium. The patient with the severely valgus knee holds the stick in the ipsilateral hand for the same biomechanical reasons (Fig 6.7).

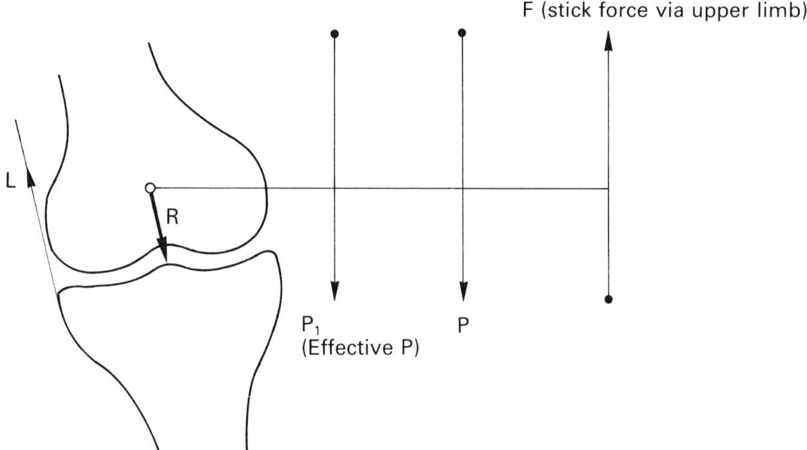

Figure 6.7 Use of walking stick. P, body weight force; P_1, body weight force (effective); F, stick force; R, resultant force; L, lateral musculo-ligamentous stay force

Dynamic appraisal

It has been pointed out that static analysis in the coronal plane can be misleading. In the gait analysis section of one study (Johnson, Leitl and Waugh, 1980), it was shown that despite clinical valgus deformity, the medial plateau, in a significant number of cases, continued to carry greater than 50% of the load transfer through the knee (Fig. 6.8). For a normal 7° valgus configuration of the knee the load is predominantly carried medially. Conversely though, as would have been expected, the lateral compartment load is reduced in clinically varus knees.

The discrepancy can be explained by the floor reaction during the stance phase of gait which produces an additional horizontal vector which is directed medially and which shifts the vertical loading toward the medial side. Consequently, the division of load between the lateral and medial tibio-femoral compartments will be more medial than predicted by static analysis (Fig. 6.9).

The meniscus and force distribution

In relation to weight (force) distribution, the menisci have a significant function for any flexion position. The area of weight-bearing in a meniscally deficient knee is approximately 50% of that where the meniscus remains intact (Maquet, 1976). The contact stresses are inevitably higher on articular

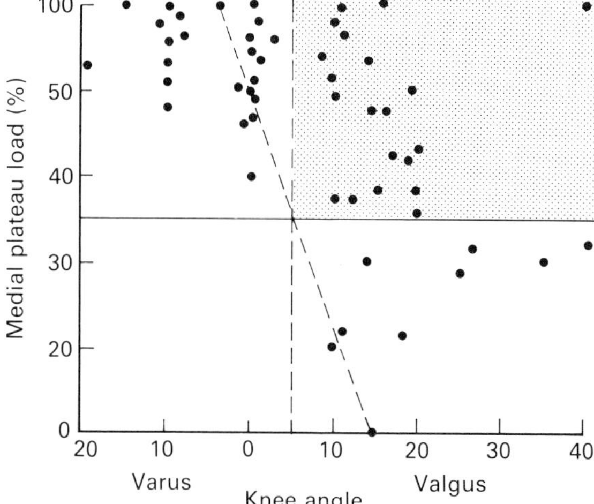

Figure 6.8 Load carried by medial tibial plateau. Results of gait analysis: 20 of the 52 joints lie in the shaded area. (From Johnson, Leitl and Waugh, 1980, by kind permission of *Journal of Bone and Joint Surgery*.)

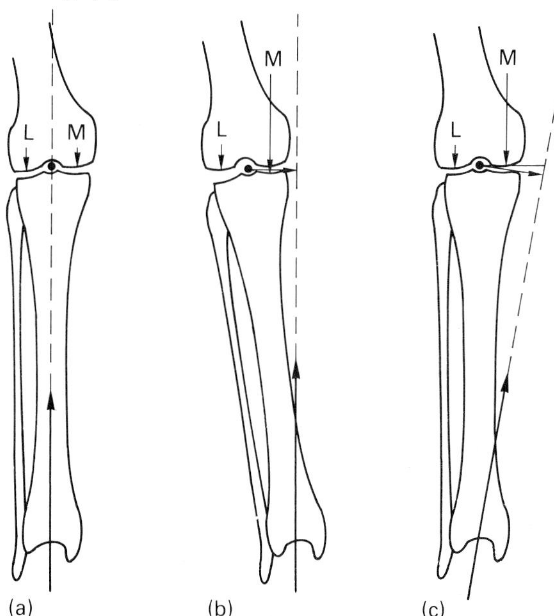

Figure 6.9 Vectors of floor reaction. (a) Normal knee, static; (b) varus knee, static; (c) normal knee, dynamic. (From Johnson, Leitl and Waugh, 1980, by kind permission of *Journal of Bone and Joint Surgery*.)

cartilage where the menisci are absent. The menisci function to produce broader conforming contact surfaces throughout the range of flexion. The effect of partial menisectomy on weight bearing or load transmission with an intact rim of meniscus, in experimental models, is to reduce the load carried by the medial meniscus by approximately 50% (for the medial meniscectomy) and 33% by lateral meniscus (for lateral meniscectomy) (Hargreaves and Seedhom, 1979). The load is assumed to be taken by the articular cartilage of the respective compartments over a smaller contact area. The effect of conservative partial meniscectomy is intermediate with respect to total meniscectomy.

Forces in the sagittal plane

The convention for analysis in the sagittal plane is with the knee slightly flexed. The body weight, acting from the centre of gravity runs forward of the hip, behind the knee and forward of the ankle joint. The resultant forces acting about the hip, ankle and calf produce a flexion moment about the knee. For equilibrium to exist, an anterior extension force is needed. This is the patella tendon force. The resultant of the flexion and patella tendon forces is a tibio-femoral compressive force which crosses the axis of flexion of the joint (Fig 6.10).

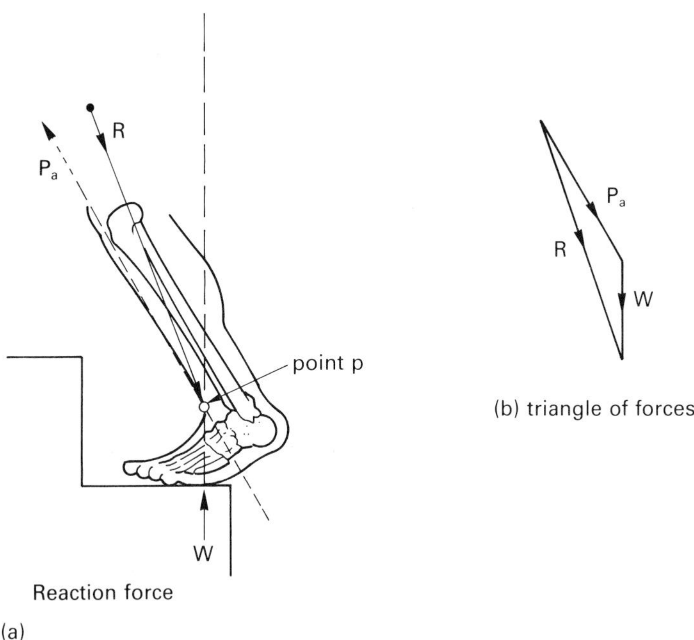

Figure 6.10 Tibio-femoral joint force. P_a, patella tendon force; W, 1 × body weight force; R, resultant tibio-femoral force. (Adapted from Franckel, 1971, Biomechanics of the knee. *Orthopaedic Clinics of North America*, **12,** by kind permission of the publishers.)

The trunk position for a given knee flexion angle, as well as the angle of flexion, is a determinant of the patella tendon force and joint force because of the manner in which the body weight force is distributed as a flexion lever arm.

The body weight, patella tendon force and joint force must intersect at a point to be in equilibrium. The joint force (resultant force) must act via the contact surface. The values of body weight force vector (magnitude and direction), patella tendon force direction and tibio-femoral joint force direction (resultant force) are known. From these values the magnitude of patella tendon force and tibio-femoral joint force (resultant force) can be calculated for any given angle of knee flexion (Franckel, 1971). The tibio-femoral joint force increases with increasing flexion angle. The tibio-femoral joint force is directly proportional to body weight, and in multiples thereof, but rarely exceeds eight times body weight even in extreme flexion (Denham and Bishop, 1978).

Patello-femoral joint forces

The force developed in the patellar tendon is balanced by the force developed by the quadriceps muscle (Fig. 6.11) (Maquet, 1976).

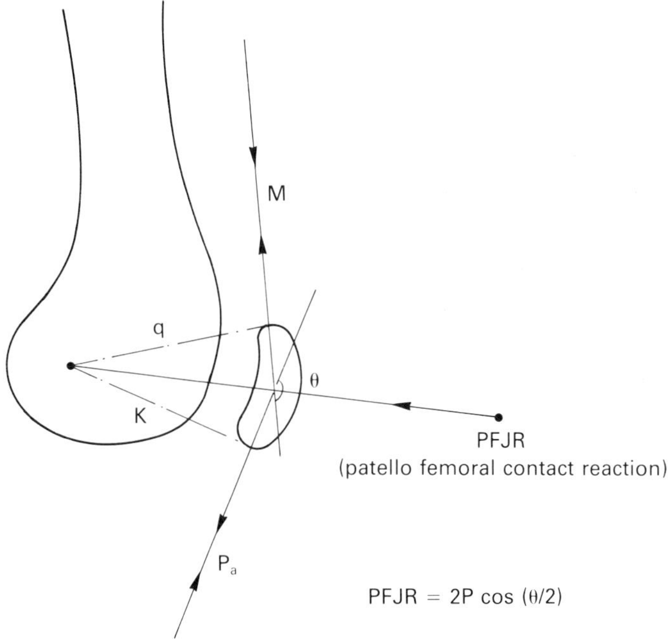

Figure 6.11 Patello-femoral forces. M, quadriceps muscle/tendon force; P_a, patella tendon force; angle between forces M and P_a; PFJR, patello-femoral joint reaction force

Formulae with reference to Figure 6.11,

$M \times q = P_a \times K$ where
M = quadriceps force; P_a = patella tendon force;
q = perpendicular distance from the centre of curvature of patellar path from vector M;
K = perpendicular distance from the centre of curvature of patellar path from vector P_a.

If the patella is assumed to be a frictionless pulley, then the patella tendon force is equal to the quadriceps muscle force. This assumption has been disputed (Denham and Bishop, 1978).

The patello-femoral joint reaction force, i.e. the compressive force on the patello-femoral joint is described by the equation:

PFJR force = $2P_a$ (cosθ/2) (Fig. 6.11)

where θ is equal to the angle between the quadriceps tendon and patellar tendon force vectors.

The resultant patello-femoral force increases with increasing patellar tendon force (and quadriceps muscle/tendon force), and with increasing flexion. Subjects walking with a stance phase flexion angle of 9–10° have a patello-femoral joint force of 0.5 times body weight. For stair climbing activities with a stance phase flexion of 60°, there is a patello-femoral joint force of approximately 3.5 times body weight and for deep knee bends at 130° of flexion, the patello-femoral joint force is approximately eight times body weight (Franckel and Burstein, 1970). In maximal flexion positions, the patello-femoral joint forces can even exceed those in the tibio-femoral joint. It is hardly surprising that stair climbing, dismounting and knee bend exercises tend to aggravate patello-femoral joint problems.

Patello-femoral contact

Goodfellow, Hungerford and Zindel (1976) have shown, with a loaded experimental model, that there is a change in the pattern of contact area between the femur and patella through the range of knee motion. At 20° of flexion the patella engages the trochlea of the femur and the contact zone between the two is at the lower pole of the patella, to either side of the median ridge. At 45°, the contact zone is more proximal about the median ridge and at 90° of flexion the contact area involves the proximal pole of the patella and the lower margins of the trochlea notch of the femur.

At 135° of flexion, the patella bears weight along the margins and on a small area medial to the medial ridge of the patella, the odd facet. In that position, the quadriceps expansions are also making contact with the condylar surfaces. This has the effect of broadening the contact surface.

Resisted extension of the knee

If the effect of body weight is eliminated and a load is applied to the foot for the knee to undergo active extension, then the quadriceps force required is approximately 10 times the applied load. For example, a 1 kg applied load requires a 10 kg quadriceps force (Franckel, 1971). Consequently, the tibio-femoral joint force is increased in direct proportion to the quadriceps force developed. This force is still greater if the rate of extension is increased, to overcome the mass moment of inertia. While the quadriceps force is greatly increased by a factor of ten in these resisted quadriceps exercises from 90° to 0°, there is also a critical zone of contact from 50° to 30° where extension against resistance produces high, non-physiological, patello-femoral forces.

These are even higher than exercises under the influence of body weight at equivalent flexion angles. It has been shown that at flexion of 90° there is an approximate contact area of 4.7 cm^2 compared with 2 cm^2 at a flexion angle of 30° (Hungerford and Barry, 1979).

Consequently, through the range from 50° to 30° of flexion, there is a smaller contact area. With resisted quadriceps exercises, there is a requirement for an increasing quadriceps force as the knee approaches extension, whereas the opposite applies with a requirement for increasing quadriceps force with increasing flexion, under the influence of body weight. Therefore the three factors,

1 Increased quadriceps force (10 × applied load)
2 Diminished contact area
3 *Increasing* quadriceps force requirement approaching the mid-flexed position, from 90° flexion

are responsible for the non-physiological excess contact forces of resisted quadriceps exercises in the range from 50° to 30° (Hungerford and Barry, 1979) (Fig. 6.12).

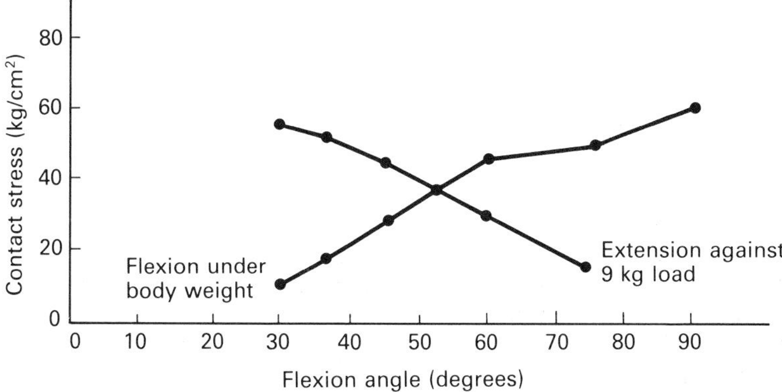

Figure 6.12 Patello-femoral contact stress. (From Hungerford and Barry, 1979, Biomechanics of the patello-femoral joint. *Clinical Orthopaedics and Related Research,* **144,** by kind permission of the publishers.)

Knee joint motion

The knee joint glides and rotates and is not a simple hinge joint. The radius of curvature of the femoral condyles is not circular but longer and flatter forward on the distal surface. The average range of motion is 0° (full extension) or a few degrees of hyperextension through to 140° (full flexion). The instantaneous centre of curvature moves and describes a cam-shaped curve (Fig. 6.13). The instantaneous centre of curvature can be found by connecting the perpendiculars to successive fixed points through the range of motion by superimposing X-ray images (Franckel, 1971). The centre of curvature, so described, optimizes gliding motion or movement at a tangent

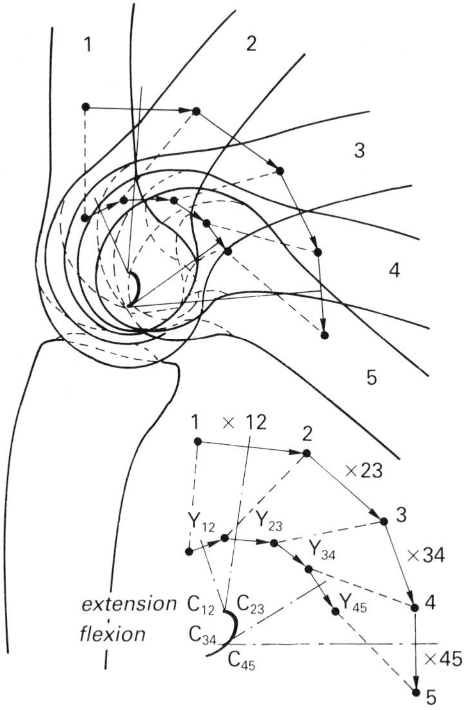

Figure 6.13 Curved path of centrode. (Adapted from Franckel *et al.*, 1971, by kind permission of *Journal of Bone and Joint Surgery*.)

to the joint surface for all flexion/extension positions. It has been shown that structural knee derangements, such as meniscal injuries, can displace the instantaneous centre of rotation at one or more points along its normal curved path. Motion about that altered point cannot be accommodated without ligament stretching and/or compressive loading to the articular surfaces (Franckel, Burstein and Brooks, 1971). In the acute phase, this may help to account for the inability to achieve full extension of the knee seen in bucket handle meniscal tears and anterior cruciate ligaments injuries. In the longer term, this phenomenon may help to explain the higher incidence of post-traumatic degenerative change that accompanies internal derangements of the knee ligaments and menisci. The range of knee rotation in the horizontal plane is 6–25° with a mean of 13° of rotational motion (Franckel, 1971). Maximal external rotation occurs with full extension just prior to heel strike.

The knee flexes to a maximum of 18–20° during the stance phase and flexes to a maximum of 75° during the swing phase of quiet walking. It has been shown that a range of less than 75° flexion will produce a visible limp during quiet walking (Franckel, 1971). The range of flexion required in weight-bearing postures is maximal for activities of crouching, sitting and negotiation of stairs. For running, the maximum flexion in stance phase is 30° and the swing phase flexion requirement is greater than 90°. During normal walking, the tibio-femoral joint force is at a peak during the late stance phase when it is approximately 3.5 times body weight, and there is a strong gastrocnemius contraction just prior to toe off. There are two smaller peaks during the stance phase which are associated with heel strike and then a quadriceps contraction at mid stance to prevent further knee bend (Franckel, 1971). The figure of 3.5 times body weight for tibio-femoral joint force underscores the importance of dynamic assessment in the sagittal plane as opposed to static analysis in the coronal plane with the knee fully extended.

The Q-angle

This is the angle formed by the directions of actions of the quadriceps muscle and the patella tendon. The angle can be measured by a line connecting the anterior iliac spine (rectus femoris origin) to the centre of the patella and a second line from that point to the tibial tubercle. The normal value is 15° and a value of greater than 20° is regarded as abnormal. The angle is measured in two dimensions but it represents a three-dimensional structure, being determined by the soft tissue and bony architecture of the hips, thigh and knee and including the rotational structure of the long bones. (Fig. 6.14).

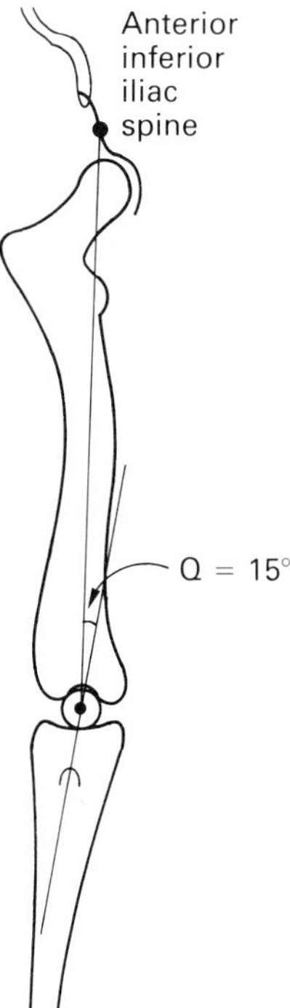

Figure 6.14 Q angle

References

Denham, R.A. and Bishop, R.E.D. (1978) Mechanics of the knee and problems in reconstructive surgery. *J. Bone and Joint Surg.*, **60B**, 345.

Franckel, V.H. (1971) Biomechanics of the knee. *Orthopaedic Clinics of North America,* **2,** 175–190.

Franckel, V.H. and Burstein, A.H. (1970) *Orthopaedic Biomechanics.* Philadelphia; Lea and Febiger.

Franckel, V.H., Burstein, A.H. and Brooks, D.B. (1971) Biomechanics of internal derangement of the knee. *Journal of Bone and Joint Surgery,* **53A,** 945–962.

Goodfellow, J., Hungerford, D.S. and Zindel, M. (1976) Patello-femoral joint mechanics and pathology. *Journal of Bone and Joint Surgery,* **58B,** 287–299.

Hargreaves, D.J. and Seedhom, B.B. (1979) On the 'bucket-handle' tear: partial or total meniscectomy? A quantitative study. *Journal of Bone and Joint Surgery,* **61B,** 381.

Hungerford, D.S. and Barry, M. (1979) Biomechanics of the patello-femoral joint. *Clinical Orthopaedics and Related Research*, **144,** 9–15.

Johnson, F., Leitl, S. and Waugh, W. (1980) The distribution of load across the knee. *Journal of Bone and Joint Surgery,* **62B,** 346–349.

Maquet, P. (1976) *Biomechanics of the knee.* New York: Springer-Verlag.

Chapter 7

Clean air theatres and the bacteriological basis for their design

Michael Kurer

The best surgery can be frustrated by sepsis. The deposition of microorganisms into the wound is followed by colonisation, destruction of tissue and toxaemia. Pasteur, in the mid-19th century, launched bacteriology into the modern world with his demonstration of the ubiquity of living organisms and their role in putrefactive and disease processes. Lister, in the UK, realized the implications for surgical practice and acted on them. He is remembered today chiefly for his use of phenol as a spray to 'destroy the life of the "floating particles"' which he realized later he was quite incapable of doing. It should also be remembered that he disinfected sutures, swabs and cut short the hair around the site of incisions and applied disinfectant to the skin of his own hands and of the patient. In 1890 in an address to the British Medical Association he said, 'we may dispense with antiseptic washing and irrigation, provided always that we can trust ourselves and our assistants to avoid the introduction into the wound of septic defilement'. (Baron, 1909). The problem of the ubiquity of microorganisms remains and in particular how to minimize the inoculation of the surgical wound.

Source of microorganisms

We must first discuss the source of infective microorganisms. The pathogenic species may originate from the patient or the operating room staff or from the outside. If from the outside, they may be carried into the room by persons or on inanimate objects (Ayliffe *et al.*, 1969). The human remains the major source of microorganisms and multiplication occurs in the upper respiratory tract, the gut, vagina and on the skin. Each of these sites has its own characteristic flora and potential for colonization and contamination. In the upper respiratory tract streptococci and staphylococci are the common organisms. These are dispersed by aerosol, principally via the saliva (Lidwell, 1974). Staphylococci are common inhabitants of the nose. There is little or no direct dispersal from this site but the organism does survive well on the skin which may become contaminated from the primary site and is passed on from there.

The skin is by far the most significant proximal source for infection. It is regularly contaminated with *Staphylococcus aureus*, in the case of nasal carriers. In addition, the same organism may colonize moist areas such as the axillae and especially the skin of the perineal area which often results in heavy contamination of the skin in general. The most obvious way which bacteria enter a surgical wound is by introduction into it of some contaminated object. Appropriate sterilization procedures effectively carried out will ensure that this is a very rare event. The hands of the surgeon and the assistants however, cannot be sterilized and, although supposedly impermeable gloves are worn, the frequency of punctures is high. Preoperative treatment of the skin of the hands is therefore required. The patient's own skin carries both its own resident microflora and contaminating organisms either from other regions of the body or from extraneous sources. The hands of the operating team cannot be sterilized, although efficient disinfecting treatment will greatly reduce the numbers of bacteria on the surface. Puncturing or cutting may transfer some of the residual organisms into the tissues beneath.

There are various reports dealing with the source of *Staph. aureus* contamination. One report (Bengtsson, Humbraeus and Laurell, 1979), reviewed infections in 1000 patients and showed that 27 occurred from a carrier state in the patient whereas 18 came from an extraneous carrier of *Staph. aureus* and 13 were from a non-carrier. Furthermore, Burke performed lavage on 50 wounds before closure and found that all contained bacteria, 46 contained a mean of five strains of *Staph. aureus* and phage typing of the staphylococci established the potential sources of origin as follows: 68% from the air above the operation wound, 50% from a carrier site on the patient, and 20% from the hands or nasopharynx of the surgical team (Burke, 1963). Obviously, some strains were isolated from more than one source, but the 68% of strains classified as airborne were found in no other situation. It is the problem of airborne contamination that is the basis of all ventilation projects.

Site of theatres

Theatres should be sited as far as practicable from sources of contamination within the hospital. Ideally they should be some distance from the wards and away from hospital thoroughfares (Fig. 7.1).

Ventilation systems

The function of a ventilating system in an operating room is as follows:

1 To supply heated, humidified contamination-free air to the operating room.
2 To introduce this air into the room so that it removes contaminants liberated there.
3 To prevent the entry of air from adjacent contaminated areas.

The efficiency of ventilation systems can be measured simply in two ways. First, the number of changes of air per hour and second, the number of microorganism-carrying particles present per cubic metre of the air supply. We will not concern ourselves with the temperature and humidity functions as these have little effect on wound infection.

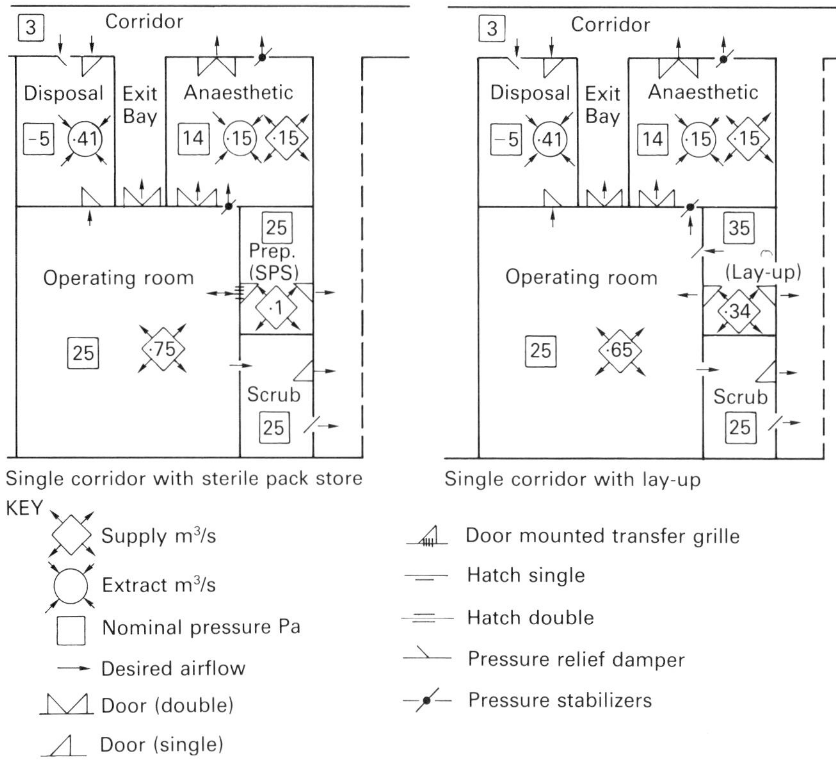

Figure 7.1 Theatre design

Plenum turbulent air flow system

The commonest and the simplest form of ventilation system present in the majority of operating departments around the country is a plenum turbulent air flow system (Fig. 7.2). Plenum is the opposite of vaccum and the system works by the provision of filtered air which is injected into the operating theatre at a higher pressure than outside thus dislodging any in the theatre and preventing the ingress of gases from outside. Air at roof level is drawn by means of fans through a series of filters. It is then humidified, cooled or heated and forced through ducts into the operating suite through high level diffusers fitted into the walls or ceiling. The air pressure in the operating suites should always be greater than that in the rest of the department and, in particular, air should move from cleaner areas to dirtier areas rather than vice versa. The highest pressure of air should always be in the preparation room where sterile trolleys are laid out, for example 35 Pa, followed by the operating theatre itself (25 Pa) scrub-up and anaesthesia room (14 Pa) and exit bay (3 Pa). The utility and disposal area should have a negative pressure (−5 Pa), in this way air circulates from clean to non-clean areas rather than vice versa (DHSS, 1983). Extraction of air from the preparation room and operating theatres is not mechanically aided. Air from the operating theatre flows through the small gaps around closed doors and also spills via balanced pressure relief flap valves set in the walls or lower part of doors. These flaps

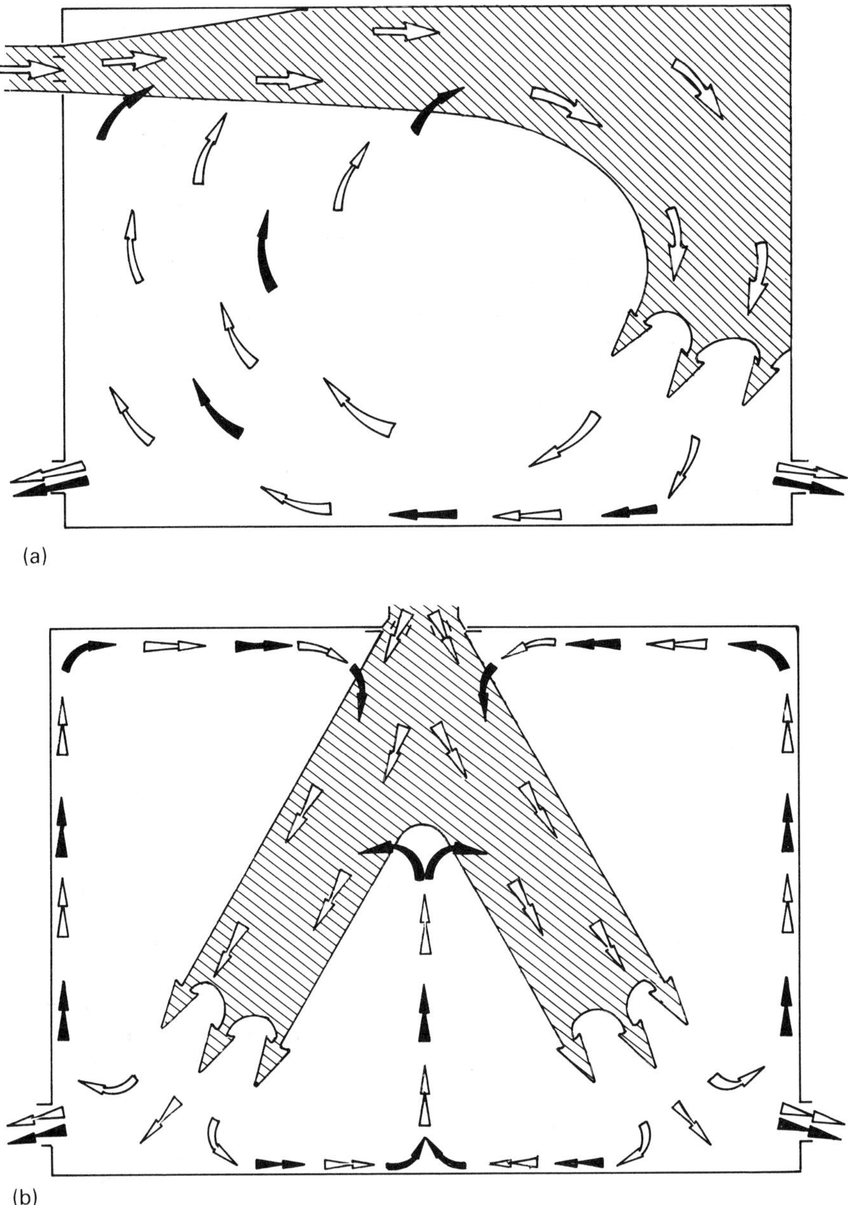

Figure 7.2 Airflow patterns in turbulent airflow systems. (a) With conventional side wall diffuser; (b) with conventional ceiling diffuser. (From Howorth, 1980, by kind permission of *Journal of Environmental Engineering*.)

are hinged and weighted so that the air is allowed to pass in one direction only. Any tendency for back flow of air from less clean air will cause them to close.

It is relatively easy to maintain an outward flow of air through the gaps around closed doors, the difficulties arise when turbulence is created by a person passing through a doorway or when a door is left open. This passage of gases will also be influenced by the difference in temperature between the two rooms. A standard doorway 2 metres high and 1.4 metres wide will allow a transit of about 0.19 m^3 per second where the temperature differential is 1°C. If the temperature difference increases to 2°C the volume of air transferred could be 0.24 m^3 per second. If several doors in an operating theatre remain open during operations this may seriously interfere with the outward air flow pattern. The total air flow supplied to an average operating theatre should be between 0.75 and 2.01 m^3 per second providing between 20 and 30 changes of air per hour.

Maintenance of these ventilation devices is important, filters must be checked regularly and there have been worries raised about Pseudomonas and Legionella contaminating cooling coils, water reservoirs and humidifying apparatus. There have been no operating theatre contracted cases of legionella reported but, nevertheless, this bacterium must be borne in mind. Humidification should preferably be by the injection of clean, dry steam either from the hospital boiler plant or a dedicated supply. Furthermore, regular bacterial sampling should be carried out.

The system described does satisfy points 1 and 3 mentioned in the aims of a ventilation system. However, it does not satisfy number 2, namely to introduce air into the room so that it removes the contaminants liberated there. We must first look in more detail at general contamination going on in the operating theatre.

Convection air currents

At rest the body emits convection currents and the physical effort of surgery gives rise to at least twice this heat emission. With a small team, the team heat output above the operating table will give rise to a convection current and a continual column of air will rise above the operating table with a velocity of 10 to 20 metres per minute. The heat of the operating lamp will accelerate this movement. The warm ascending air meets with the filtered air introduced through inlets in the roof, sweeps downwards at the the perimeter of the room across to the table and upwards again (Fig. 7.3). Sixty per cent of the clean air clings close to the wall and passes out of the room via the exit valves without contributing to the removal of contaminants. Turbulent flow and downward displacement systems are incapable of opposing this convection pattern of air movement. Indeed, jets of turbulent flow may actually accentuate the problem. The system is further provoked by the movement of staff, who not only produce local eddies and alter the pattern of air flow, but increase the number of airborne contaminates by mobilising those safely lodged on the floor.

Emission of particles

The number of particles emitted by a human being is very much greater than is generally realized. These particles carry bacteria. It is only those with a diameter of greater than 100 μm that can be seen by the unaided eye. Each person in the operating room is emitting at least 10 000 bacteria-carrying

Figure 7.3 Airflow around a conventional ceiling diffuser. (From Howorth, 1980, by kind permission of *Journal of Environmental Engineering*.)

particles per minute when at rest and with increased activity this can rise to 50 000 per minute. (Bethune *et al.*, 1965) (Fig. 7.4). It is these particles, which once airborne, are thought to contribute greatly to wound infection. There are some people (9% males and 6% females) who are 'high dispersers', with rates three to four times those of normal people. The majority of the particles come from below the waist and leak into the room via the openings of trousers and skirts (Whyte, Vesley and Hodgson, 1976). The increased production with activity was first thought to be caused by the abrasive effect of the clothes on the skin's surface. More recent work has shown that particles are shed from the skin surface of the naked subject with equal facility. Presumably it is the flexing of the skin's surface that is important. Once shed the movement of a particle in the operating room is completely random; radioactive tracer studies have shown it may make as many as eight to twelve traverses of the room before discharge.

Laminar flow system of ventilation

'Laminar Flow' is something of a misnomer. It was first used as a verbal shorthand to describe the parallel linear pattern of air flow that is characteristic of this low velocity, low turbulence form of movement. Since, however, it does not fulfil the aerodynamic conditions for genuine laminar flow this name should be avoided. 'Unidirectional' or 'linear' air flow have been suggested as alternatives and the latter seems to evoke the actual pattern of movement in parallel lines. This form of ventilation was evolved

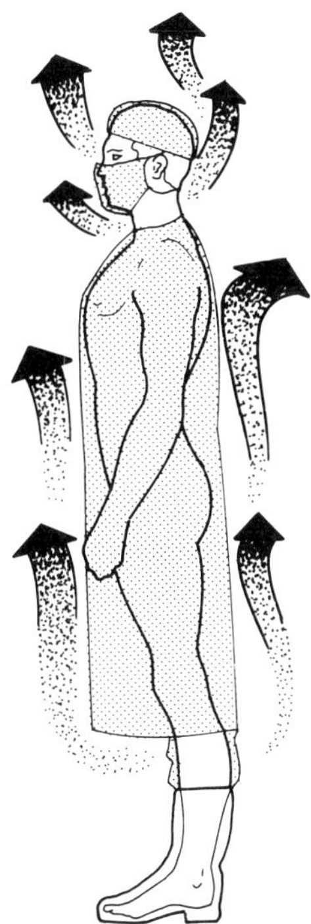

Figure 7.4 Convection currents around the surgeon. (From Howorth, 1980, by kind permission of *Journal of Environmental Engineering.*)

in the laboratories of the Sandia Corporation of Alberquerque, New Mexico in 1960. It is a nuclear weapon research and development laboratory and the team of ventilation engineers was led by a physicist, Dr Willis Whitfield, (Whitfield, 1962). The success of mechanical parts on NASA space missions has in part been attributed to their production in clean air environments. The fundamental difference between the existing systems of ventilation and the laminar/linear flow ventilation can be summarized as follows. In the conventional approach emphasis is placed on limiting the amount of contamination liberated into the air by controlling the number and movements of the people in the room. Contamination is removed by cleaning and maintenance. By contrast the laminar/linear flow system utilizes the highly filtered and conditioned air brought into the room *towards* the critical work, through a filter bank comprising an entire wall or an entire ceiling and making only a single transit over any given area of the room. As the entire volume of air in the room moves in one mass at a velocity of 30 metres per minute, convection currents due to heat or movement are abolished and the re-entrainment of particles into the operative field is impossible. When a solid object is encountered the air flows round the object and the laminar/linear pattern is distorted only in the immediate area of the object (Fig. 7.5). All critical work is thus performed in a uniform flow of clean air. Contaminants are flushed out as soon as they are liberated and migration to other areas is prevented by the striations of air flow.

This latter point has remained a bone of contention and, in the original

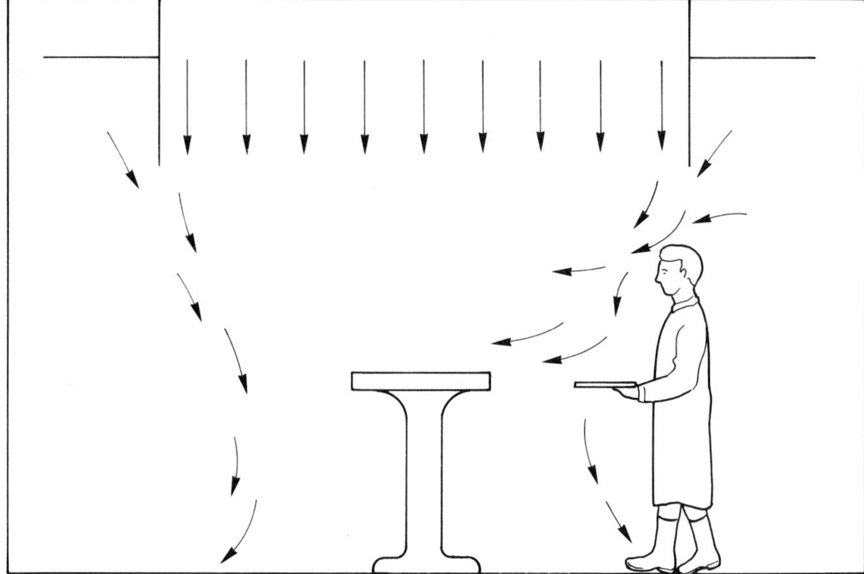

Figure 7.5 Vertical laminar flow. (From Howorth, 1980, by kind permission of *Journal of Environmental Engineering.*)

design, walls from the ceiling and down to within 1 m of the floor were required in order to prevent extraneous air being sucked into the field (Charnley, 1964). Vertical flow systems are preferred because of their superior performance but, in an existing operating theatre, horizontal systems may be the only alternative. These are unfortunately much less effective. These systems may give up to 500 to 600 air changes per hour in the operating area. The hanging side panels provided considerable restriction to the surgeon, the other members of the surgical team, the anaesthetist and the positioning of instruments. Furthermore, the introduction of larger objects of equipment like image intensifiers often lead to the side panels being omitted and so rendering the system ineffective. Horizontal flow, though initially felt to be quite useful, has since been shown to be ineffective in a trial from the Hospital for Special Surgery (Salvati *et al.*, 1982). It was found that in many cases, wound sepsis was significantly higher with horizontal unidirectional flow when compared with conventional air conditioning systems in an adjacent operating room.

Allander has further suggested a centrally located diffuser panel with a high velocity air curtain all round its periphery. The central air speed is 0.25 metres per second from a 3 metre diffuser. Around the periphery of the diffuser there is an aperture from which an air curtain is discharged at 0.6 metres per second. Trials carried out in Canada show that there was 11 times the level of wound sepsis with the Allander compared with the Charnley/Howorth system in an adjacent operating room.

The latest and most modern design is the exponential ceiling flow, ceiling diffuser (Howorth, 1980). This is the only air flow pattern which is able to provide a truly clean zone for surgery and which will overcome any challenge from outside the clean zone without necessity of walls, panels or other obstructions. The principle is based on the mouth of the trumpet facing downwards. A trumpet is an exponential horn. Because this air flow pattern is substantially trumpet shaped it is known as exponential flow. This is abbreviated to 'Ex flow'. The Ex flow sterile air pattern has a higher velocity and pressure in the centre of the clean zone than at the periphery. This ensures that it dominates the surroundings and therefore peripheral entrainment cannot occur. This system is more effective than the linear flow and only 100 room changes per hour are required. The air flow pattern of this shape has the facility of moving the sterile air essentially downwards over the wound and the surgical team. The air then curves progressively outwards towards the periphery of the clean zone and beyond. The flow of sterile air is always moving radially outwards away from the surgical team (Fig. 7.6). Anyone or anything approaching must move against the outwards flow of sterile air. No airborne contaminants which may challenge it from outside can enter this clean zone.

Other measures

Because of the large number of bacteria-carrying particles being emitted by people, it is essential that sterile instruments be exposed from their wrappings only in the zone which is free from airborne contamination, otherwise the benefits of sterilization will be greatly diminished. In order that the trays and packs of instruments which have been sterilized do not become contaminated while they are held in store, they should be kept on shelves in open fronted laminar flow cabinets. From the rear face of these cabinets a positive flow of sterile air moves horizontally over the shelves and everything which is on them. The re-circulated air in the room provided by these cabinets, helps to remove the body bacterial emissions of the personnel who work there. This, consequently, will greatly reduce, by dilution, the general level of airborne bacteria-carrying particles which are present in sterile instrument stores and preparation rooms (Howorth, 1980).

Contamination by anaesthetic gases

Even with a good air-conditioning system, there are many zones in which the air is contaminated by the emission of anaesthetic gases to levels which are unacceptably high and which may contaminate the wound with respiratory bacteria (Fig. 7.7). Accordingly a high volume, low pressure scavenging system is recommended (Hatch, Miles and Wagstaff, 1980). Drapes should also be arranged to exclude the patient's head and the anaesthetist from the operative field.

Body exhaust systems

Since body emissions are the main source of wound sepsis, those who are actually working in the clean zone should have their body emissions removed. The majority of the emissions come from below the neck. For the removal of these a gown should envelop the body from the top of the head

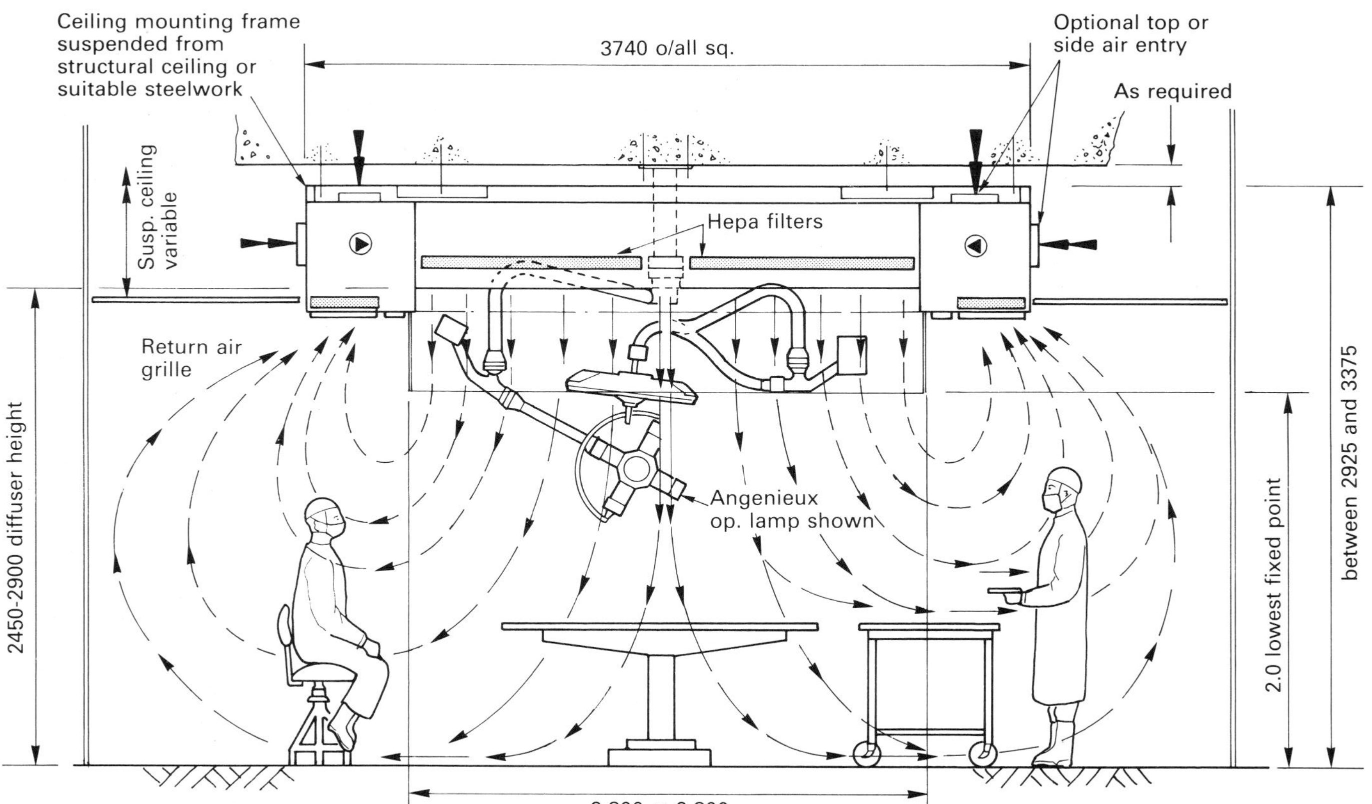

Figure 7.6 The Howorth exflow 90. The non-entrainment airflow pattern of the Howorth exflow 90 which provides an enlarged ultra clean zone without any side wall restrictions. (From Howorth, 1980, by kind permission of *Journal of Environmental Engineering*.)

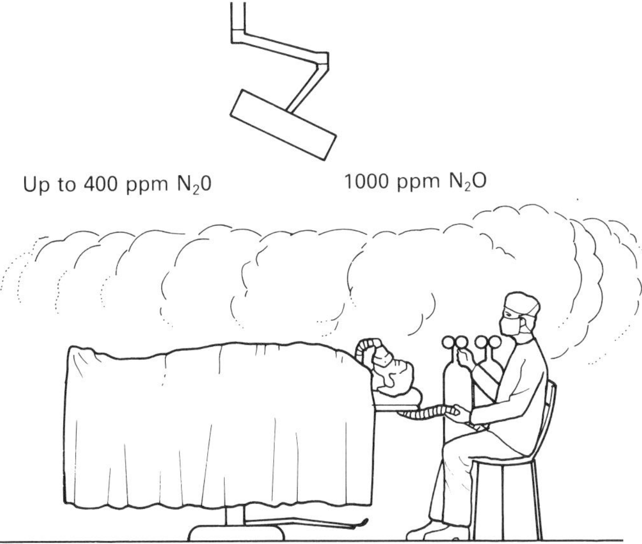

Figure 7.7 Contamination by anaesthetic gases. (From Howorth, 1980, by kind permission of *Journal of Environmental Engineering.*)

down to not more than 45cm from the floor. Air is drawn from under the gown at a volume which is sufficient to create a negative pressure within the gown. At any apertures air will move inwards in order to satisfy the demands of the exhaust system. The material from which the gown is made must have low permeability to air and be impermeable to water. It must not allow the passage of bacteria, either by air or by the wicking or capillary actions of liquids. There are four basic types of body exhaust gown.

1 Charnley gown, which is of the pull over type. It integrates the Charnley plastic face mask and visor and it is from either side of this mask that the exhaust system draws off air in order to create a negative pressure inside the gown and down to within about 45 cm from the floor. In this way all body emissions are removed, also cooling and oxygenation are improved and an audio system can be added. However, it is not possible to look into an arthroscope or a microscope wearing such a gown.
2 In a modification for the USA, a head piece which is formed as a transparent sphere can rest upon the shoulders of the wearer so that the head can move within the sphere.
3 The Mandarin gown, which allows the use of optical devices. With this attire the gown terminates at the neck with a close fitting mandarin collar. It is made from the same type of material as the Charnley gown and is worn with a conventional nose and mouth mask. Air is exhausted from under the gown by a necklace type of body exhaust tubing fitting over the shoulders and down to the waist where it is tied with string. Because around 90% of the body bacterial emissions come from below the neck, this is an effective compromise.
4 The fourth type of body exhaust gown envelops the wearer from the top

of the head down to 45 cm (18 inches) from the floor. It has been designed to overcome all objects and restrictions of previous gowns. With this gown there is an aperture in the gown at the eyes and across the bridge of the nose so that spectacles as well as other optical devices may be used in the normal way.

Results

	Bacteria carrying particles per m^3
Exhaust ventilated theatre	750–1000
Conventionally air conditioned plenum type ventilation	400–500
Horizontal air flow	34
Downward air flow	8
Downward air flow with the surgical team wearing a body exhaust system	0.63

Thus the total reduction in bacteria carrying particles measured at the wound site from an ordinary operating theatre to the maximum is 650 times. This is irrespective of any antibiotic prophylaxis (MRC Study, Lidwell *et al.*, 1982).

The major study on the actual effect of reducing bacterial counts on clinical wound infection rate was performed by Lidwell *et al* and reported in 1982. I shall go into some detail on this as it is often quoted. The aim of the study was to perform a trial on three different operation groups. The first group were patients whose operations were performed in a standard turbulent plenum ventilation supply. The second group were operations performed in an ultra-clean air system and the third were those performed in an ultra-clean air system with the surgeons wearing body exhaust suits. This large study was started up in 1974 by the Medical Research Council. To give a reasonable chance (90%) of establishing a difference between sepsis rates of 2% in a control series had a 95% confidence level. It was estimated that about 2500 would be necessary in each group. Ultra-clean air was arbitrarily defined as containing less than 10 bacteria carrying particles per m^3. In total 7500 operations were needed for significant results and therefore a multicentre trial was organized. Patients were followed up for at least 12 months and 19 different hospitals participated; 11 in England, four in Scotland and four in Sweden. Hip and knee replacements were performed and in all 8136 operations were entered of which 8055 could actually be used because of data infringements or other problems. There were 6781 operations on the hip and 1274 on the knee. The average follow up was 2 years. In addition, bacteria in air samples were taken at operation and the wounds were also irrigated in a percentage of patients to monitor bacterial counts. Three thousand wound wash samples were collected in all as well as a large number of dummy samples. Results of this trial were as follows:

1 The bacteria washings from operation wounds in ultra-clean air rooms were always small and in the cleanest air conditions the numbers were the same as dummy samples.
2 There were 86 cases of sepsis or probable sepsis and the bacteria isolated were *Staphylococcus aureus*. Of these infections 63 out of 4133 (1.5%) were from the control group and 23 out of 3922 (0.6%) were from the ultra-clean group, giving a ratio of 2.6 times with 95% confidence limits, which was highly significant.
3 Combining the operations with wearing body exhaust systems gave a ratio of 4.5 times with a 95% confidence limit.
4 There were some antibiotics used in the series, and attempts have been made to draw conclusions from their use. It appeared that the overall incidence of sepsis in patients given antibiotic prophylaxis was less than one third of patients who did not receive prophylaxis and that this effect was cumulative with the effects of ventilation.
5 Finally there was one hospital who had a particularly high rate of infection and it was noted in this hospital that the highest counts of airborne bacteria were present in their conventionally ventilated operating rooms.

A further study by Lidwell in 1984 has shown that where at least 100 joint replacement operations are done each year, the cost of installation and maintenance of an ultra-clean ventilation system is more than off-set by the financial savings through prevention of infection even when prophylactic antibiotics are used (Selwyn, 1986).

Conclusion

In conclusion, our understanding of the mechanisms of wound infection have been somewhat enhanced by the work reported in this review. It is quite clear that ventilation systems do have an effect on reducing the bacterial count in theatres and, therefore, wound infections. However, much of this work has been superseded by the use of prophylactic antibiotics. Wound infection rates are now so small it has become hard to implicate ventilation systems when so many other factors may play a part. Nevertheless, reducing bacterial counts can only be of benefit and accordingly should be the surgeon's aim. With the advent of HIV infection and the possible inoculation of surgeons and staff, body exhaust systems may offer a double advantage to both patient and staff and perhaps should be re-evaluated.

References

Ayliffe, S.A., Babb, J.R., Collins, B.J. and Lowbury, E.J.L. (1969) Transfer areas and clean zones in operating suites. *Journal of Hygiene*, **67:** 417–25.

Baron, J. (1909) The collected papers (2 vols) volume 2, p. 37 and 170: 93 & 336. Clarenden Press, Oxford.

Bengtsson, S., Humbraeus, A. and Laurell, G. (1979) Wound infection after surgery in a modern operating suite. *Journal of Hygiene*, **83,** 40–57.

Bethune, D.W., Blowers, R., Parker, M.T. and Parke, E. (1965) Factors affecting the dispersal of bacteria carrying particles in theatre personnel. *Lancet*, i, 480.

Burke, J.F. (1963) Analysis of staphylococci from 50 wounds isolated at surgery. *Annals of Surgery*, 158–898.
Charnley, J. (1964) A sterile air operating theatre enclosure. *British Journal of Surgery*, **55**, 195–202.
Department of Health and Social Security (1983). *Ventilation of Operating Departments*. A design guide. London: HMSO.
Elmslie, J.A.W. (1966) Staphylococcus dispersion by surgical staff. *Lancet*, i, 660.
Hatch, D.J., Miles, R. and Wagstaff, M. (1980) An anaesthetic scavenging system for paediatric and adult use. *Anesthesia,* **35:** 496–99.
Howorth, F.H. (1980) Airflow patterns in the operating theatre. *Journal of the Environmental Engineers*, June 29th.
Howorth, F.H. (1984) The air in the operating theatre. In: *The Design and Use of Operating Theatres*. (Johnston, I.D.A. and Hunter, A.R. eds,) London, Arnold.
Lidwell, O.M. (1974). Aerial dispersal of micro-organisms for the human respiratory tract. In: *The Normal Microbial Flora of Man*. (Skinner, F.A. and Carr, J.G. eds.) Academic Press London, pp. 135–54.
Lidwell, O.M., Lowbury, E.J.L., White, W., Blowers, R., Stanley, S.J. and Lowe, D. (1982) Effect of ultra-clean air in operating rooms on deep sepsis in the joint after total hip or knee replacement: a randomised study. *British Medical Journal*, **285:** 10–14.
Salvaty, E.A., Robinson, R.P., Zeeno, S.M. *et al*. (1982) Infection rates after 3,175 total hip and total knee replacements performed with and without horizontal unidirectional filtered airflow system. *Journal of Bone and Joint Surgery*, **64A:** 523–35.
Selwyn, R.C., (1986) Cost-effectiveness of clean air operating Theatres. *British Medical Journal,* i, 897–898.
Whyte, W. (1969). Bacteriological aspects of air-conditioning plants. *Journal of Hygiene*, **66**, 567–84.
Whyte, W., Vesley, D. and Hodgson, R. (1976) Bacterial dispersion in relation to operating room clothing. *Journal of Hygiene*, **76:** 367–78.
Whitfield, W.J. (1962) Sandia Corporation Publications 1962 SC-4673 (RK).

Chapter 8

Indications and methods for skin and tissue cover in trauma

Boyd Goldie

Classification

There have been many attempts to classify soft tissue trauma. Although the systems are useful in allowing comparison between different series of patients, they do not better a clear description of a wound. The significance of a 1cm skin puncture over the subcutaneous border of the tibia does not have the same significance as a similar wound associated with a fractured shaft of femur.

The Tscherne classification has the advantage that it considers *all* injuries, whether open or closed (Table 8.1). Included are those injuries with severe soft tissue damage that, although closed, need to be treated with as much respect as those that are open.

Table 8.1 The Tscherne classification for soft tissue trauma (Tscherne and Gotzen, 1984)

Contusion	*Skin*	
	Closed	*Open*
None	Fr. C O	
Small	Fr. C I	Fr. O I
Circumscribed	Fr C. II	Fr. O II
Extensive	Fr C. III	Fr. O III
Amputation		Fr. O IV

Fr. C O *Grade O closed fracture:* soft tissue damage is absent or negligible. The fracture is caused by indirect violence and has a simple configuration.

Fr. C I *Grade I closed fracture:* there is a superficial abrasion or contusion caused by fragment pressure from within. The fracture itself is of a mild to moderately severe configuration.

Fr. C II *Grade I closed fracture:* there is a deep contaminated abrasion associated with localized skin or muscle contusion from direct trauma. Generally there has been direct violence producing a moderately severe to severe fracture configuration.

Fr. C III *Grade III closed fracture:* the skin is extensively contused or crushed, and muscle damage may be severe. Other criteria for this category are subcutaneous avulsions, decompensated compartment syndrome, and rupture of a major blood vessel associated with a closed fracture. The fracture configuration is severe or comminuted.

Fr. O I *Grade I open fracture:* open wound with little or no skin contusion. Generally the fracture is of a mild configuration.

Fr. O II *Grade II open fracture:* Open wound with circumscribed skin and soft tissue contusions and moderate contamination. The severity of the fracture is variable.

Fr. O III *Grade III open fracture:* open, heavily contaminated wound with extensive soft tissue destruction, often with associated vascular and nerve lesion. Any open fracture with ischaemia and extensive comminution.

Fr. O IV *Grade IV open fracture:* total or subtotal amputation.

The grading system from Lange (1985) is widely used for open fractures.

Grade I A skin wound from within, usually less than 1 cm or no skin contusion.

Grade II A skin wound of more than 1 cm, with skin and soft tissue contusion, but no loss of muscle or bone. A small wound over a major muscle mass should be considered a grade II open fracture.

Grade III A large, severe, open wound with extensive skin and subcutaneous contusion, muscle crush or loss, and severe periosteal stripping. This grade is subdivided into three subgrades:
- a a wound associated with severe bone loss, muscle loss, nerve or tendon injury.
- b a wound associated with an arterial injury.
- c a traumatic amputation.

Gustilo and Anderson's original (1976) classification used three grades for open fractures of long bones:

Type I An open fracture with a wound less than 1 cm and clean.

Type II An open fracture with a laceration more than 1 cm long without extensive soft-tissue damage, flaps or avulsions.

Type III Either an open segmental fracture, an open fracture with extensive soft-tissue damage, or a traumatic amputation.

In 1987 Gustilo subdivided the Type III injuries into the following groups:

Type IIIA Open fractures have adequate soft-tissue coverage of a fractured bone despite extensive soft-tissue laceration or flaps, or they result from high energy trauma irrespective of the size of the wound. This includes segmental fractures or severely comminuted fractures from the high energy trauma irrespective of the size of the wound.

Type IIIB Open fractures have extensive soft tissue loss with periosteal stripping and bony exposure. This is usually associated with massive contamination.

Type IIIC Open fractures are associated with arterial injury requiring repair, irrespective of degree of soft-tissue injury.

Primary treatment

Decision making

The factors that need to be considered in the decision making for treatment of an open injury include general factors:

- Patient's age
- General health of the patient
- Presence of other injuries

The local factors that should be considered are:

- The extent of the soft tissue injury
- The time between injury and definitive treatment
- The fracture configuration
- Presence of injury to major structures, especially vascular.

Immediate treatment

Cultures should be taken in the casualty department prior to the wound being covered with a sterile dressing as quickly as possible. Once covered, the wound should remain undisturbed until in the operating theatre for definitive treatment. Tscherne has found a significant increase in the infection rate of open fractures not covered by a sterile dressing from the accident scene to the operating theatre. The use of a polaroid camera in the casualty department may reduce the temptation for multiple unnecessary viewings of the wound.

The patient should be given a third generation cephalosporin antibiotic (Gustilo, *et al.*, 1984). Anti-tetanus prophylaxis must also be given if appropriate.

Limb salvage

In an open fracture with a vascular injury, limb salvage may not be appropriate. It is the functional result and quality of life that are of prime importance. Sixty per cent of all open tibial fractures with vascular injury eventually require amputation and the decision to amputate should be made early and at a senior level (Pozo, 1990).

Cleansing

The skin itself must be cleaned of all visible dirt and then prepared for surgery in the standard fashion. The agent used for preparing the skin must not damage the wound or have systemic effects. Many commercial surgical preparation solutions containing iodophors and hexachlorophane are not safe for use in a wound and potentiate the risk of infection. Experimentally contaminated wounds treated with these solutions develop more infections than wounds irrigated with 0.9% saline.

The wound can be cleaned by scrubbing and/or irrigation. A scrubbing brush will remove embedded debris from most wounds. Irrigation of the wound is best performed with physiological saline. There is still some argument whether or not high or low pressure irrigation should be used. High pressure irrigation may produce further trauma, but low pressure irrigation is less effective at removing small particles (bacteria).

Debridement

'Débrider' is the French verb to unbridle, or in surgery, to incise.

Debridement is the most important part of the acute management of the contaminated wound. Contaminated tissue and devitalized tissue must be removed. The need for devitalized tissue to be removed is because:

1 It acts as a culture medium.
2 It inhibits leucocyte phagocytosis and bacterial killing.
3 The anaerobic environment within the devitalized tissue limits leucocyte function.

Crushed skin edges are excised conservatively. The amount of degloving of the skin should be noted and care must be taken not to increase it. Skin that is crushed and degloved should be excised. The difficulty is in deciding the limit of the skin excision. One should be reasonably conservative, as further excision can be performed after 48 hours.

The indicators that skin is viable are
1 The appearance of dermal bleeding from the cut skin edge.
2 The distribution in the skin of fluoroscein injected intravenously and observed under an ultra-violet light (McGrouther and Sully, 1980).

The fluoroscein test has the disadvantages that its interpretation is difficult and the dye may provoke a hypersensitivity reaction.

A useful diagnostic and therapeutic technique is to suture an avulsed skin flap back in position and take a split thickness skin graft from the whole flap. Dermal capillary bleeding on the flap surface indicates which part of the flap is viable and, in addition, the skin graft can be used immediately or stored (Ziv, 1988).

Dead and devitalized muscle must be excised. The indicators that the muscle is viable are:

1 Colour.
2 Twitching on mechanical stimulation.
3 Bleeding.

Consideration of how soft tissue cover is to be subsequently achieved should never limit the extent of the debridement.

If compartment syndrome is possible, the deep fascia over muscle should be widely released.

For all grade III open fractures, an experienced plastic surgeon, interested in trauma, should be in theatre. Ideally he or she should be present at the initial debridement or at the first wound inspection at 48 hours.

Fracture stabilization

Fracture stabilization is necessary to promote soft tissue healing. The choice usually lies between internal fixation or external fixation. It should be emphasized that stable internal fixation is *not* contraindicated in open fractures and does not necessarily result in an increase in the rate of wound infection (Chapman, 1979).

Implant selection may be summarized as follows:

1 External fixation – for grade III wounds associated with severe comminution or bone loss in either the metaphysis or diaphysis.
2 Open reduction and internal fixation – for open metaphyseal fractures with or without joint involvement, using interfragmentary screw and plates. Open reduction and internal fixation is also suitable for diaphyseal fractures, which can be anatomically reduced because of their configuration, in a grade I or II wound.
3 Combined fixation – minimal internal fixation with an external frame for comminuted fractures in grade II wounds, where lag screws are used to improve the stability of the fracture.
4 Intramedullary nails – are nowadays used in the primary treatment of Gustillo grade I and II open fractures. They may be used inserted after the initial debridement or secondarily after the wound has been assessed and no sepsis found, usually on the 5th to 14th day following trauma. The use of intramedullary nails in the Gustillo grade III open fractures is still a matter for debate.

Initial wound closure or cover

In considering the various options for obtaining soft tissue cover, the correct option is that which provides reliable skin cover rapidly, and leaves the limb in a permanently functional and stable condition.

Primary closure is occasionally possible in grade I wounds. If a wound is closed primarily, the following should be avoided:

1 Tension.
2 Dead space.
3 Relaxing incisions.
4 Primary flaps.

One can never be certain, however, that devitalized tissue has not been left in the wound and closure may precipitate sepsis. It is safest, therefore, to leave all wounds open. Any extensions of the wound may be closed without tension. Sensitive structures, such as joints, nerves and tendons should have soft tissue cover.

The wound should be covered with a dressing that prevents drying and contains an antiseptic. Paraffin gauze soaked in antiseptic is an effective dressing.

Secondary wound care

The initial inspection of the wound should be on the second or third day after trauma. It is at this time that the decision about the best method of skin cover should be made.

If necessary, serial debridement should be performed. The aim is to gain bacteriological control. Bone coverage should only be performed when all the tissues are clean and healthy. Remember that living bone rarely desiccates.

Healing by secondary intent

If the wound is small and clean, or if the wound can only be partly closed by secondary suture, it may be left open to heal by secondary intent.

Secondary suture (delayed primary closure)

This is indicated if:

1 There has been no skin loss.
2 There is no sign of sepsis.

Secondary suture of part of the wound may reduce the area to be covered by skin grafts.

Secondary closure under excessive tension in the zone of trauma may precipitate skin necrosis.

Free skin grafting

A split thickness skin graft will take on any surface capable of sprouting capillary loops; viable muscle, subcutaneous fat, loose areolar tissue and periosteum.

For a successful take, the wound must be clean, with healthy granulation tissue.

A split thickness skin graft will not take reliably onto; bare bone, cartliage, tendon, joint capsule or implants.

The advantages of a split thickness skin graft are:
- Easy to perform
- Can be extended by meshing to cover a large area
- Permits wound observation
- Minimal donor morbidity.

The major disadvantage is that it will not take onto the surfaces described and it may make secondary procedures difficult.

Local skin flaps

Local skin flaps can be used for small wounds, but their application is limited. They have to be designed and handled with great care to avoid edge necrosis and traumatized skin cannot be used in the flap.

Fasciocutaneous flaps can be raised with a greater length to width ratio than skin alone.

1 Transposition flap. The flap is designed so that the diagonal length is greater than the distance it will have to reach across the defect. A split skin graft is needed to cover the resultant defect (Fig. 8.1).
2 Sliding transposition flap. The flap slides proximally, producing a dog-ear which is ignored.

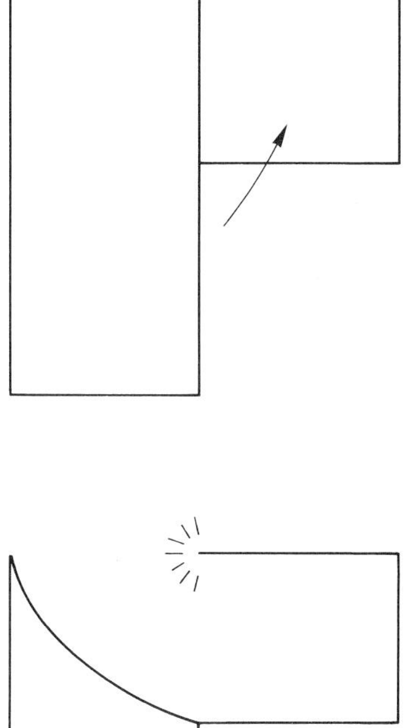

Figure 8.1 The design of a transposition flap (top) and after transposition (bottom). A length–width ratio such as this is only safe when deep fascia is included with the skin. The diagonal length of the flap must exceed the distance it has to reach when transposed. The donor defect requires a split skin graft. (From Evans, 1988, *Current Orthopaedics,* **2,** 3–8, by kind permission of the publishers, Churchill Livingstone.)

Local muscle flaps

Muscles adjacent to the tibia may be divided distally and brought across bare bone. Although anterior compartment muscles could be used, they are not because a functional deficit would remain.

The indications for a local muscle flap include: exposed bone, denuded tendons or open joint.

The advantages of a local muscle flap include: provides its own blood supply, allows secondary procedures and limited morbility.

The disadvantages are: the muscle to be transposed may be in the zone of trauma and they are difficult for distal wounds or large defects.

The choice of muscle flap include;

1 Flexor hallucis longus – useful since muscle belly reaches the ankle.
2 Soleus or either head of gastrocnemius – upper tibial cover.
3 Distally based soleus flap – unreliable.

Local myocutaneous flaps

The gastrocnemius myocutaneous flap has the advantage that the skin component can be extended along the gastrocnemius tendon and can carry a fasciocutaneous component even further. Therefore the flap can cover lower and larger defects than muscle alone. Medial or lateral heads of gastrocnemius may be used. While handling a myocutaneous flap it is best temporarily to suture the skin to the muscle to avoid shearing the perforating vessels.

This flap is useful in covering defects over the upper and middle thirds of the tibia.

A bipedicled flap maintains both upper and lower connections of the flap. It is useful in covering long narrow defects. The incision on the far side of the flap has to be much longer than the defect, to allow advancement without tension, and muscle and fascia are included in the flap (Fig. 8.2).

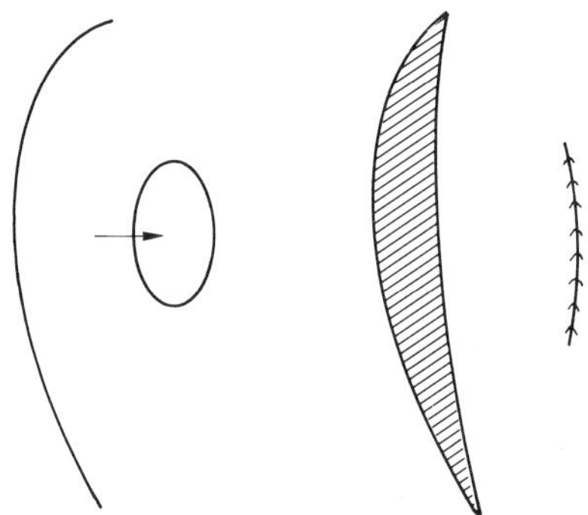

Figure 8.2 A design for a bipedicled flap to close an elliptical defect. The length of the flap greatly exceeds the length of the defect, and the incision on the far edge of the flap has to curve round above and below the defect so that movement of the flap does not cause a line of tension across the flap, which should move easily in its new position. This should be a myocutaneous flap, otherwise the blood supply at its critical leading edge is unreliable. (From Evans, 1988, *Current Orthopaedics,* **2,** 3–8, by kind permission of the publishers, Churchill Livingstone.)

Distant pedicle flap

A distant pedicle flap is one where skin is transferred to another part of the body on a pedicle containing the vascular supply. The length of the vascular pedicle determines the distance that the flap can be moved.

The indications for a distant pedicle flap include a well- defined defect.

The advantages are: provides blood supply, an axial pattern and a high success rate.

The disadvantages are that it is often bulky, the limb may need to be attached to distant donor and it may be saprophytic after division of pedicle.

Examples of a pedicle flap include:

- Groin flap – based on the superficial circumflex iliac artery
- Radial forearm flap – based on the radial artery
- Deltopectoral flap – based on the internal mammary artery
- Dorsalis pedis flap – based on the dorsalis pedis artery. Can be used to cover defects around the ankle.
- Instep island flap – based on the medial plantar vessels and nerves. It is useful for heel defects.

Free flaps

Previously, when the skin coverage was required from a distant site, conventional flaps were pedicled from one extremity to another or tubed from donor to recipient in multiple stages. Nowadays, most major soft tissue defects are reconstructed using free tissue transfer. A free flap may be defined as a composite tissue transfer which is transferred in one stage to a distant recipient site and revascularized via a microvascular anastomosis (Daniel, 1978).

A free flap transfer consists of three steps:

1 Identification of recipient vessels with the artery possessing pounding pulsatile flow and one or more veins offering adequate drainage.
2 Isolation of a viable island flap based on only those vessels which will be anastomosed at the recipient site.
3 Transfer of the flap to the recipient site with revascularisation through microvascular anastomoses.

The advantages of free flaps include:

- Coverage of larger defects than local or regional flaps.
- Provision of bulk to a defect.
- Transfer of sensate skin, muscle or bone.
- Enhanced blood supply to injured area.
- Donor sites may be closed directly.
- Allows for secondary procedures.
- May be performed immediately.
- Have permanent vascular pedicle.
- Fit defect precisely.
- Permit elevation and early mobilization of the limb.
- Avoid additional damage to an already injured limb, that is inflicted by a local flap.

The disadvantages of free flaps include:

- Difficult, long and tedious operation.
- Require specialized surgical team.
- Failure of a flap is complete and disastrous.
- Need for careful post-operative monitoring.

The prerequisites for wound preparation for a free flap are:

- A stable skeleton.
- No contamination.
- No tissue with a compromised blood supply.
- As flat a surface as possible.

In choosing a free flap the following factors should be considered:

- The general health of the patient.
- The comfort of the patient.
- The cosmetic demands of the patient.
- Diabetes or peripheral vascular disease may be contraindications to a free flap.
- Flaps from the chest wall should be avoided in patients with respiratory problems.
- If the patient may need prolonged bed rest, flaps from the back should be avoided.
- Scars over the upper arm, forearm or scapular should be avoided in girls.
- The size of the defect.
- The thickness of the defect.
- The need for bone, tendon, nerve, artery and sensation will influence choice of flap.
- The skill of the surgeon.
- The experience of the team.

Results

In most series, a 90+% flap survival is the norm. Free flaps have a shorter immobilization period and shorter in-patient stay when compared will other forms of skin flap coverage, such as the cross-leg flap.

The review by Khouri and Shaw (1989) of free flaps for the lower extremity highlights some interesting points.

- They performed 283 flaps for trauma patients, in 10 years. The majority were performed for acute large open wounds.
- Nearly half of their flaps used the latissimus dorsi flap. The parascapular flap and the rectus abdominis flap were the next most commonly used.
- Forty-seven per cent were needed for mid to distal leg and 33% were used to cover the ankle and foot.
- They achieved a 91% success rate and most of the failures were in the mid to distal leg where the recipient vessels may have been traumatized.
- Forty five (15%) of their flaps thrombosed postoperatively, but 20 of these were saved by a re-exploration.
- Twenty five flaps were performed on the day of injury. These were for patients with the most severe trauma with exposed vital structures which required immediate coverage. Only 8% of these failed.
- The last 100 flaps had half the rate of failure and re-exploration than their first 100 flaps.

The commonly used free tissue transfers are given below.

Flap: Scalp/forehead flap
Type: Cutaneous
Major vessel: Superficial temporal artery
Length: 2–3 cm (scalp), 0–10 cm (forehead)
Diameter: 1–1.8 mm
Size of flap: 15 cm × 3 cm (scalp), 7 cm × 5 cm (forehead)
Application: Head reconstruction
Advantages: Primary closure – scalp
Disadvantages: Donor defect forehead

Flap: Scapular
Reference: dos Santos, (1984) *Plastic Reconstructive Surgery*, **73,** 599–603
Type: Cutaneous, osteocutaneous
Major vessel: Circumflex scapular artery
Length: 6–7 cm
Diameter: 2 mm
Size of flap: 10 × 24 cm
Application: Large area
Advantages: Thin skin, no donor problems, easy access
Disadvantages: Positioning in lateral or prone positioning while flap harvested

Flap: Parascapular
Type: Cutaneous
Major vessel: Circumflex scapular
Length: 7–10 cm
Diameter: 2.5–3.5 mm
Application: Large area
Advantages: Thin skin, no donor problems, easy access
Disadvantages: Positioning

Flap: Latissimus dorsi
Reference: Chaikhouni (1981) *Journal of Trauma*, **21,** 398–402
Maxwell (1980) *Plastic and Reconstructive Surgery*, **65:** 686–692
Type: Musculocutaneous
Muscular
Major vessel: Thoracodorsal artery
Length: 8–14 cm
Diameter: 1.5–3 mm
Size of flap: 40 × 20 cm
Application: Large soft tissue defect
Advantages: Large consistent vessels, long pedicle
Disadvantages: Bulky, disability with crutches

Preparing the latissimus dorsi muscle (Green, 1988)

The entire muscle or part of it can be transferred with or without a cutaneous cover. The donor site is cosmetically unacceptable when covered by split skin, if it cannot be closed directly. This limits the width of the skin flap to a maximum of 10 cm. For larger soft tissue defects, muscle alone is taken which is then covered with split skin. This allows for better contouring of the graft to the defect.

The vascular supply is primarily by the thoracodorsal artery which enters the muscle 8–12 cm from its insertion. Two venae commitantes combine to form a single vein near the axillae which varies in diameter from 3.0 to 5.0 mm.

To harvest the latissimus dorsi, the patient is placed in the lateral decubitus position or the prone position.

The anterior margin of the muscle is marked by drawing a line from the anterior aspect of the posterior axillary fold to a midlateral position on the iliac crest (Fig. 8.3).

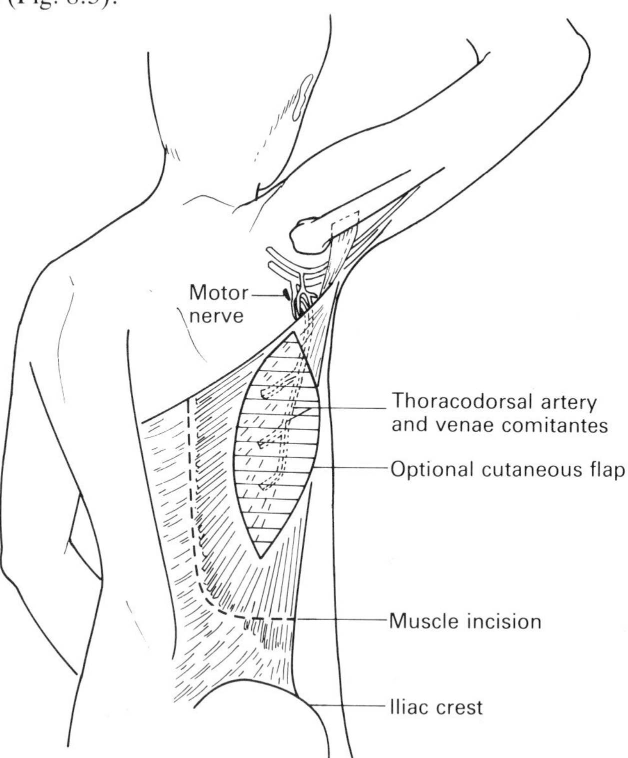

Figure 8.3 Diagrammatic representation of the anatomy of the latissimus dorsi flap. (From Green, D.P. (ed.) *Operative Hand Surgery*, 2nd edn, vol 2, p. 1223, by kind permission of the publishers, Churchill Livingstone.)

The cutaneous flap, if used, is taken from the superior and lateral portion of the muscle where the skin is particularly well vascularized.

A curving incision is begun in the axilla along the anterior border of the muscle. The anterior border of the muscle is followed down inferiorly until the required amount of muscle is exposed. The muscle is divided distally and the dissection carried upwards parrallel to the spine. The pedicle is palpable 1–2 cm from the anterior margin. The long thoracic nerve is identified and spared. The pedicle is traced back to the circumflex scapular artery to give a pedicle up to 10 cm long.

Flap:	Serratus anterior
Reference:	Takaynagi (1982) *Annals of Plastic Surgery,* **8:** 277–283
Type:	Muscular, osseous-muscular
Major vessel:	Thoracodorsal artery
Length:	15 cm
Diameter:	2–3 mm
Size of flap:	10 × 10 cm
Application:	Hand reconstruction
Advantages:	Very long vascular pedicle, well located donor site scar

Flap:	Lateral upper arm
Reference:	Katsaros (1984) *Annals of Plastic Surgery,* **12:** 489–500
Type:	Cutaneous, osteocutaneous, neurosensory
Major vessel:	Posterior radial collateral artery
Length:	6 cm
Diameter:	3mm
Size of flap:	10 × 15 cm
Application:	Hand, upper extremity
Advantages:	Innervated, fascia-fat flap
Disadvantages:	Donor defect on lateral arm

Flap:	Forearm
Type:	Fasciocutaneous, innervated fasciocutaneous, osteocutaneous
Major vessel:	Radial artery
Diameter:	2–3 mm
Size of flap:	Forearm
Application:	Hand and foot reconstruction, maxillo-facial
Advantages:	Thin flap, up to 10 cm bone
Disadvantages:	Donor site cosmesis, hairy flap

Flap:	Pectoralis major
Reference:	Ikuta (1976) *Plastic Reconstructive Surgery,* **58:** 407–411
Type:	Musculocutaneous, muscle
Major vessel:	Thoraco-abdominal
Length:	4–10 cm
Diameter:	1.5–3mm
Size of flap:	20 × 30 cm
Application:	Large soft tissue defect, head and neck reconstruction
Advantages:	Large, broad and innervated
Disadvantages:	Small durability, prior site deformity Short pedicle Cosmetically poor donor site

Flap:	Rectus abdominis
Reference:	Taylor (1984) *British Journal of Plastic Surgery,* **37,** 330–350
Type:	Musculocutaneous, muscle
Major vessel:	Inferior epigastric artery
Length:	5 cm
Diameter:	2.5 mm
Size of flap:	10 × 15 cm
Application:	Soft tissue defects intermediate between those suitable for gracilis or latissimus dorsi flaps
Advantages:	Easy to harvest, expendable
Disadvantages:	Donor site defect – hernia Cannot be used if previous surgery to the groin

Flap:	Groin
Reference:	McGregor (1983) *British Journal of Plastic Surgery,* **26:** 202–213
Type:	Cutaneous
Major vessel:	Superficial circumflex iliac artery
Length:	2–4 cm
Diameter:	0.8–3mm
Size of flap:	20 × 30 cm
Application:	Hand and forearm, lower extremity, head/ neck large soft tissue defect
Advantages:	Easy to harvest, primary closure, may include bone
Disadvantages:	Bulky Inconsistent vascular supply

Flap:	Gracilis
Reference:	La Rossa (1980) *Journal of Trauma,* **20,** 545–550
Type:	Musculocutaneous, muscular
Major vessel:	Dominant muscle artery
Length:	4–6 cm
Diameter:	1–2 mm
Size of flap:	33 × 10 cm
Application:	Innervated muscle for forearm, dorsum of foot
Advantages:	Long flap, thick, good donor site scar
Disadvantages:	Variable skin quality. Double vascular pedicle

Flap:	Gluteus
Type:	Musculocutaneous, muscle
Major vessel:	Superior gluteal
Length:	0–6 cm
Diameter:	1.5–2.5mm
Size of flap:	15 × 10 cm
Application:	Thick coverage
Advantages:	Breast reconstruction
Disadvantages:	Bulky, not expendable muscle. Short pedicle

Flap:	Tensor fascia
Type:	Musculocutaneous, muscle
Major vessel:	Tensor fascia lata artery (lateral femoral circumflex)
Length:	6–8 cm
Diameter:	2 mm
Size of flap:	10 × 30 cm
Application:	Upper and lower limb soft tissue defects
Advantages:	Size, donor defect, no functional loss, neurosensory flap
Disadvantages:	Short pedicle,variable vessel. Bulky

Flap:	Dorsalis pedis flap
Reference:	O'Brien (1973) *Australian and New Zealand Journal of Surgery,* **43:** 285–288
Type:	Skin, osteocutaneous, joint
Major vessel:	Dorsalis pedis
Length:	6–8 cm
Diameter:	6–8 cm
Size of flap:	15 × 5 cm
Application:	Upper limb
Advantages:	Thin. May be sensate
Disadvantages:	Possible donor site disability. Flap size limited to dorsum of the foot

References

Chapman, M.W. (1979) The role of early internal fixation in the management of open fractures. *Clinical Orthopaedics and related Research,* **138,** 120–131.

Daniel, R.K. (1978) Free flaps: An overview. *Clinical Orthopaedics and Related Research,* **133,** 122–131.

Evans, D.M. (1988) Principles of soft tissue cover. *Current Orthopaedics,* **2,** 3–8.

Green, D.P. (ed.) (1988) *Operative Hand Surgery*, 2nd edn, Vol. 2 Churchill Livingstone, Edinburgh, p. 1222.

Gustilo, R.B. and Anderson, J.T. (1976) Prevention of infection in the treatment of open fractures of long bones. *Journal of Bone and Joint Surgery,* **58A,** 453–458.

Gustilo, R.B., Mendoza, R.M. and Williams, D.N. (1984) Problems in the management of Type III (severe) open frac-tures: a new classification of Type III open fractures. *Journal of Trauma,* **24,** 742–6.

Gustilo, R.B. (1987) Classification of Type III (severe) open fractures relative to treatment and results. *Orthopedics,* **10,** 1781–1788.

Lange, R.H. (1985) Open tibial fractures with associated vascular injuries: prognosis for limb salvage. *Journal of Trauma,* **25,** 203.

Khouri, R.K. and Shaw, W.W. (1989) Reconstruction of the lower extremity with microvascular free flaps: A 10-year experience with 304 consecutive cases. *Journal of Trauma,* **29,** 1086–1094.

Mathes and Alpert BS. (1988) Free skin and composite flaps. In: *Operative Hand Surgery*, (Green, D.P., ed.). Churchill Livingstone, Edinburgh.

Manktelow, R.T. (1988) Free muscle transfers. In: *Operative Hand Surgery*, (Green, D.P., ed.) Churchill Livingstone, Edinburgh.

McGrouther, D.A. and Sully, L. (1980) Degloving injuries of the limbs: long term review and management based on whole body fluorescence. *British Journal of Plastic Surgery,* **3,** 9–24.

Pozo, J.L. (1990) The timing of amputation for lower limb trauma. *Journal of Bone and Joint Surgery,* **72B,** 288–292.

Tscherne, H. and Gotzen L. (1984) *Fractures with Soft Tissue Injuries.* Springer, Berlin.

Ziv, I. (1988) Split thickness skin excision in severe open fractures. *Journal of Bone and Joint Surgery,* **70B,** 23–26.

Chapter 9

The pathology of osteoporosis

Allan N. Stirrat

The incidence of fracture secondary to osteoporosis continues to increase. In the USA the prevalence of osteoporosis is estimated at 15–20 million people. There are 1.2 million fractures per year secondary to osteoporosis and over one third of females over the age of 65 years have sustained a vertebral crush fracture at some time.

In the UK in 1989, there were approximately 130 000 vertebral fractures, 40 000 distal forearm fractures and 50 000 proximal femoral fractures. The incidence of Colles' fracture rises dramatically over the age of 50 and that of femoral neck fracture over the age of 70 (Grimley Evans, 1990a). The majority of these fractures are at least partly related to the underlying presence of osteoporosis. The costs of treating these injuries are enormous and will increase steadily as the elderly population expands. Estimates from the Royal College of Physicians report (1989) on prevention and management of femoral neck fracture suggest that the incidence of proximal femoral fracture will lie between 60 000 per year at the most conservative projection and 117 000 per year, if the increasing trend continues, by 2016.

Investigation of the pathology of osteoporosis has a large part to play in the determination of aetiology and consequently the prevention of fractures.

Definition

Osteoporosis may be defined as a condition in which 'the bones are of normal dimensions but contain less bone tissue per unit volume, although the bone tissue is histologically normal' (Grimley Evans, 1990a).

Alternatively, it may be stated that 'osteoporosis is characterised by a decrease in bone mass which renders the skeleton more liable to fracture' (Kanis *et al.*, 1990).

The wide ranging term osteopenia is often used synonymously with osteoporosis. Osteopenia is a general descriptive term for any cause of loss of bone mass or increased radiographic lucency, e.g. osteomalacia, neoplastic disease and collagen disorders.

Bone turnover

Formation and destruction of bone, that is remodelling, occurs throughout life (Fig. 9.1). This process relies on two main groups of bone cells, osteoblasts and osteoclasts. The osteoblast, which is thought to be derived from the marrow stromal stem cell, is responsible for the synthesis, organization and mineralization of the extracellular matrix of bone, in particular type 1 collagen. Osteocytes are considered to be osteoblasts that have remained behind the advancing front of bone formation.

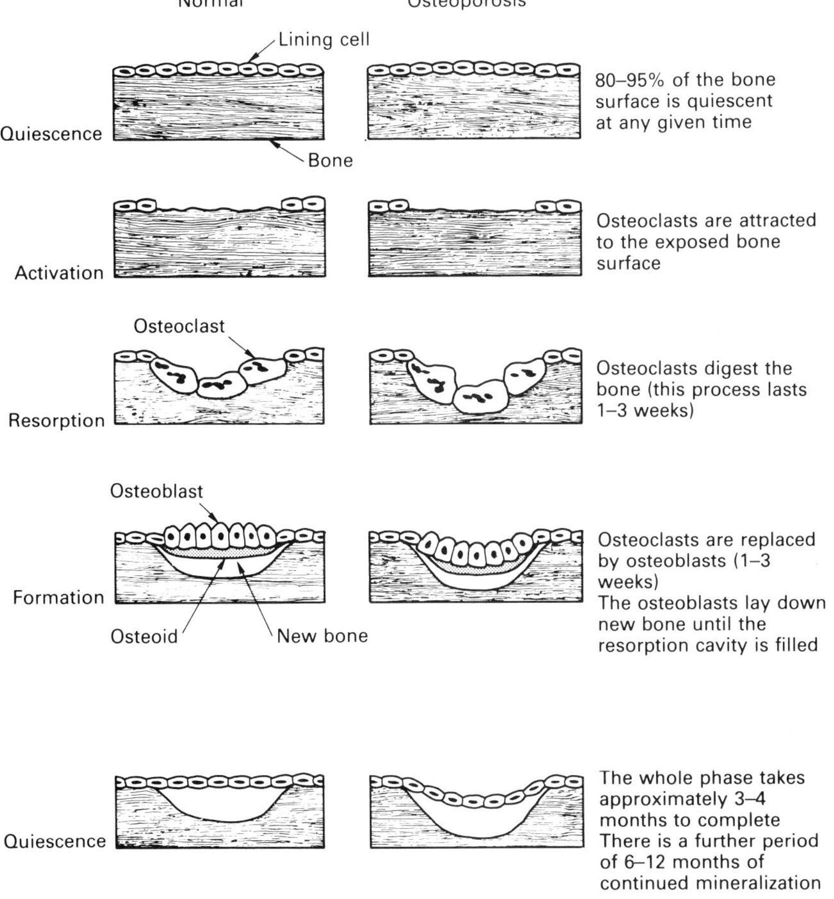

Figure 9.1 Diagram detailing phases of bone remodelling. (Redrawn with kind permission of Norwich Eaton Ltd.)

The haematopoietic stem cells give rise to the osteoclast which is principally involved in the resorptive process. These large cells contain many nuclei and display a ruffled border beneath which bone resorption takes place. This resorption depends on the generation of a local acid environment which aids degradation. The mobility of the osteoclasts plays an important part in the resorptive capacity (Russell, 1990).

Under normal circumstances, in the young patient, bone formation equals bone destruction. There is 10–30% turnover of the skeleton per year. Thus skeletal mass is maintained by the process of remodelling. This balanced state is known as coupling. A multitude of factors including cytokines (interleukins, tumour necrosis factors), growth factors (platelet derived growth factor, nerve growth factor, bone morphogenic proteins) and other peptides are involved in the control mechanisms of osteoblasts and osteoclasts. It is not known yet how these agents interact, although the synergistic and the antagonistic mechanisms are extremely complicated.

In the clinical situation it is recognized that cytokines related to malignant disease may induce the resorptive process. Oestrogen is thought to inhibit release of promoters of bone resorption and also may influence osteoblast proliferation. These mechanisms have not been defined as yet.

Classification

Osteoporosis may be divided into primary and secondary groups. A pathological distinction is drawn between postmenopausal osteoporosis (type I: accelerated phase) and senile osteoporosis (type II: slow phase) (Riggs and Melton, 1983).

Postmenopausal or accelerated osteoporosis occurs predominantly in the 5 years after onset of the menopause. Increased numbers of osteoclasts precipitate erosion and disproportionate loss of trabecular bone.

Slow phase osteoporosis commences about the age of 30 years and occurs at a steady rate in both sexes, manifest by proportionate loss of cortical and trabecular components due to impaired bone formation.

Pathology

The skeleton is composed of cortical and trabecular bone in respective proportions of 80% and 20%. Remodelling occurs predominantly on bone surfaces, the majority of which are trabecular. Trabecular bone has a high surface to volume ratio and is more prone to excessive resorption or an imbalance of remodelling. Although in the context of osteoporosis, remodelling appears a harmful process, it is essential for maintenance of bone strength through self repair. When the coupling mechanism is disrupted, 'uncoupling' erosion of the trabecular surfaces occurs. The components of this process are (1) thinning of trabeculae and (2) discontinuity and loss of trabeculae which is termed 'loss of connectedness' (Compston, 1990) (Fig. 9.2).

The second feature prevails in postmenopausal osteoporosis and is due to generation of resorption cavities which actually transect or perforate the trabeculae. this type of erosion has a devastating effect on structural strength out of proportion to the amount of material removed. It is a consequence of the increased turnover which occurs secondary to the loss of oestrogen. Osteoporosis associated with a low turnover state, e.g. the senile (type II) variety, or that induced by corticosteroids, is manifest by trabecular

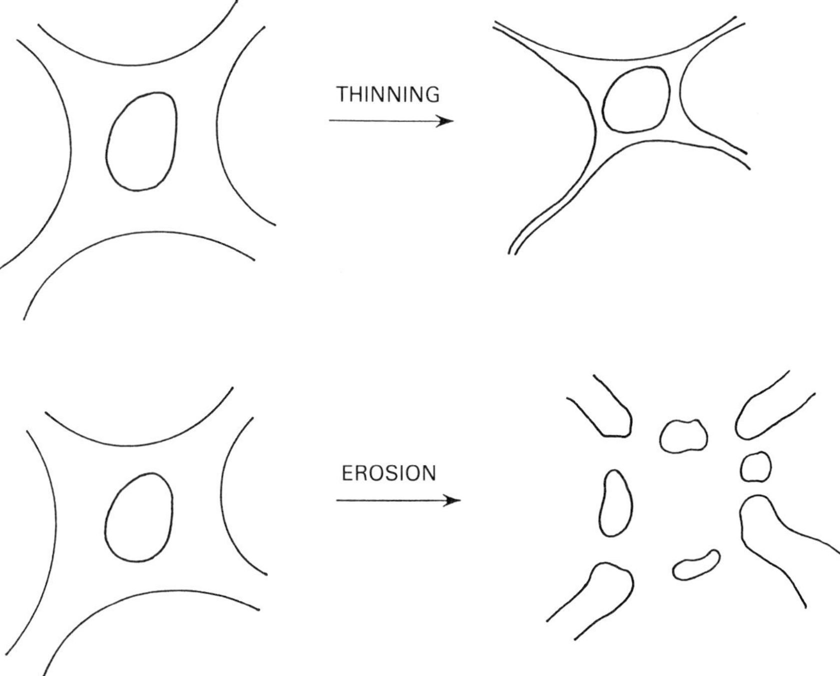

Figure 9.2 Diagrams demonstrating contrast between thinning of trabeculae as in senile osteoporosis and erosion as in postmenopausal osteoporosis

thinning only. In theory, trabecular thinning is a more reversible phenomenon than trabecular loss.

Bone resorption may be quantified by measurement of resorption cavity dimensions from iliac crest bone biopsies (Compston, 1990). Such studies suggest that the loss of connectedness of trabecular bone seen in postmenopausal women occurs because of an increase in the number of remodelling units rather than from increased resorption depth. A total of 1–5% of the total trabecular surface is involved by resorption cavities.

Multifactorial aetiology

Factors which influence bone density combine to determine whether osteoporosis occurs and at what rate it advances. Of critical importance is peak bone mass (PBM) which is normally achieved in the early 20s in both sexes. PBM is greater in the male. Failure to attain expected PBM is a strong predeterminant of osteoporosis. Determining factors may be grouped as follows:

1 Oestrogen lack.
2 Exercise.
3 Calcium intake.
4 Alcohol and medication.

5 Smoking.
6 Genetic.

Oestrogen lack

Reduction of circulating oestrogen, notably oestradiol in the perimenopausal and postmenopausal period is the most important contributory factor to osteoporosis. Fuller Albright in 1940 was the first to notice this connection. Parity is known to have a protective effect. Greater longevity also renders the female more liable to osteoporosis. Bone loss occurs in men over the age of 50 at a rate of 0.4% per annum. In women, from the age of 30 until the menopause, loss occurs at a rate of 0.75–1.00% per annum, accelerating to 2–3% through the menopause and for 5 years afterwards (Griffin, 1990). At this rate a woman may lose one third of her bone mineral by the age of 70 (Smith, 1987). This has otherwise been expressed as loss of 35% of cortical bone and 50% of trabecular bone (Riggs and Melton, 1986). Commencement of trabecular loss precedes that of cortical loss by at least 10 years. Loss of cortical bone may be measured by metacarpal radiographs and that of trabecular bone by volume to volume ratio of the iliac crest (Fig. 9.3). The incidence of fracture in bone with high trabecular content, such as the vertebrae and distal radius, increases after the menopause. Crush and wedge fractures of the vertebrae are associated with postmenopausal and senile types respectively. The rate of fracture of the proximal femur which has a greater component of cortical bone, does not accelerate until the age of 70 years.

It is important to remember that osteoporosis is one of a triad of factors implicated in proximal femoral fracture, the others being tendency to fall and incoordination (Grimley Evans, 1990b). Trabecular loss in the proximal femur has been assessed radiologically by Singh's index (Fig. 9.4).

Exercise

Conflicting evidence exists as to whether exercise is beneficial in the premenopausal phase. One American study showed higher vertebral bone density in association with high exercise level and high dietary calcium (Kanders, Dempster and Lindsay, 1988). Similar measurements in British premenopausal women revealed no difference before the menopause. However, postmenopausal women who regularly took part in weight-bearing exercise demonstrated increased bone density in the proximal femur in comparison with non-exercisers (Stevenson *et al.*, 1989). Swimming does not augment vertebral trabecular density in postmenopausal women suggesting that only weight-bearing exercise is beneficial in that group (Orwoll *et al.*, 1987). Thus exercise may reduce bone loss but cannot negate the effect of oestrogen lack. In postmenopausal women body weight is proportional to bone density in the vertebrae and the proximal femur. This may be due to increased load bearing and to the greater proportion of fat which is a potent source of oestrogen production.

Diet and Calcium

Around 27 mmol of calcium is the required intake per day in adolescents, of which 50% is absorbed. Adequate protein intake is necessary to allow

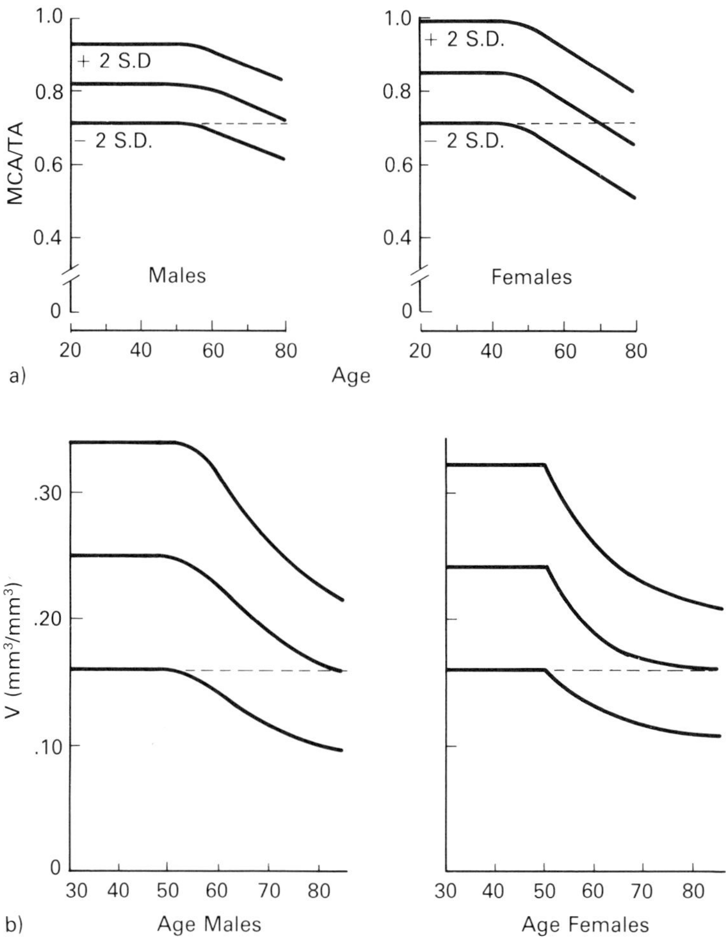

Figure 9.3 Graphs showing (a) metacarpal bone loss with age in men and women (MCA/TA: metacarpal cortical area/total area). (b) Change in iliac crest trabecular bone volume in men and women (volume to volume ratio of iliac crest trabecular bone). The interrupted lines are the lower limits of normal. (Reproduced from Nordin, B.E.C. (ed.) 1984, *Metabolic Bone Disease,* with kind permission of Churchill Livingstone.)

calcium absorption. In the growing phase the skeletal requirement of calcium is 10 mmol per day. Research evidence suggests that dietary intake of calcium during bone growth is an important factor in determining peak bone mass. The dietary deficiency of the early twentieth century may have contributed to the recent rise in incidence of proximal femoral fracture. Matkovic et al (1979), comparing two areas of Yugoslavia reported 50% lower incidence of proximal femoral fracture in the area with a relatively high calcium intake. However, it should be noted that the rate of Colles' fracture did not differ between the two areas. A recent prospective study in the USA (Holbrook, Barrett-Connor and Wingard, 1988) suggested up to 60% reduction in the risk of proximal femoral fracture in both sexes when calcium intake was greater than 19 mmol per day. High calcium intake

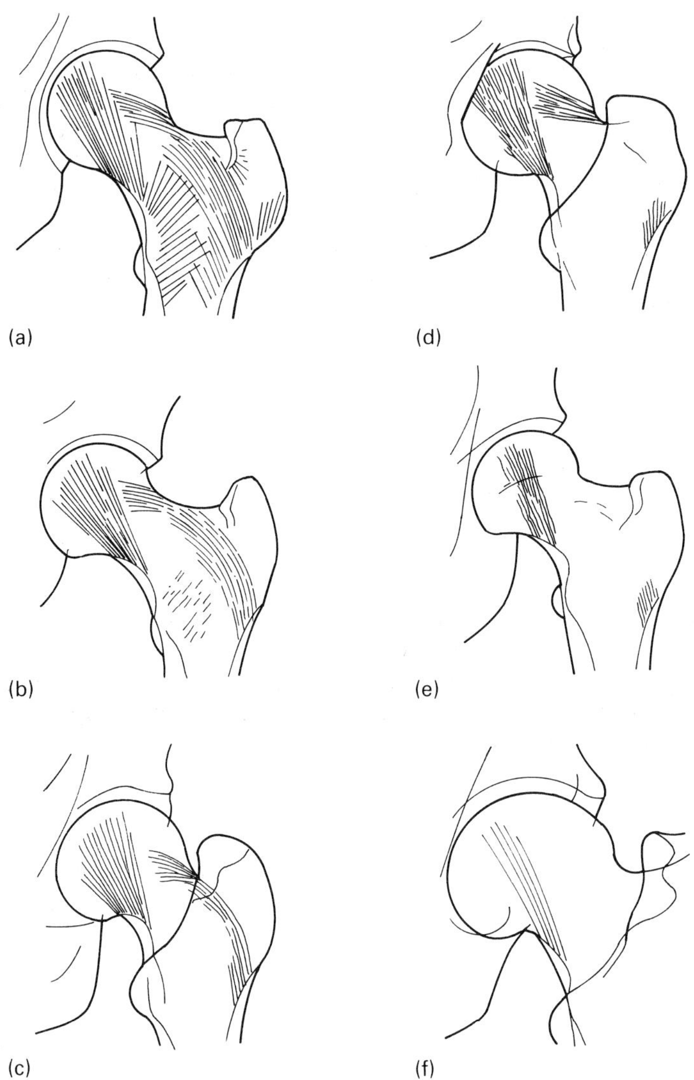

Figure 9.4 Diagrammatic summary of Singh index for assessment of osteoporosis in the proximal femur. (From Singh, M., Nagrath, A.R. and Mani, P.S. 1970, Changes in trabecular pattern at the upper end of the femur as an index of osteoporosis. *Journal of Bone and Joint Surgery*, **52A**, 457–467. Redrawn with kind permission of *Journal of Bone and Joint Surgery*.)

(2000 mg per day) has been shown to inhibit cortical loss but not that of trabecular bone after the menopause, although calcium supplementation was not as effective as oestrogen therapy (Riis, Thomsen and Christiansen, 1987). Cortical preservation may reduce the occurrence of proximal femoral fracture. Thus calcium intake plays a major part in attainment of PBM, but alone has only a minor influence on bone resorption. The consensus of research suggests a daily dietary requirement of 20–25 mmol of calcium.

After the menopause, calcium absorption is decreased due to declining cutaneous production of vitamin D and its metabolites. Lower calcium absorption has been demonstrated in osteoporotic patients suggesting that calcium supplementation should reduce the rate of bone loss. Although vitamin D increases intestinal calcium absorption, it does not limit bone loss and appears to increase postmenopausal resorption. The mechanism of calcium supplements may be pharmacological rather than physiological, in that by increasing ionized calcium, parathyroid hormone and 1,25 DHCC are suppressed.

Fluoride is known to promote bone deposition. Ansell and Lawrence (1966) showed variation in bone density according to the concentration of fluoride in the water supply between the two English towns of Watford and Leigh.

Alcohol

Alcohol makes a multifactorial contribution to osteoporosis and fracture. In excess it is associated with impaired nutrition, decreased body weight, reduced activity, increased incidence of liver disease and greater frequency of falling.

Cigarette smoking

Female smokers have lower circulating levels of oestrogen and have a tendency to earlier menopause. Reduced body weight and physical inactivity may also play a part. It has been suggested that the increasing incidence of proximal femoral fracture is related to the upsurge of smoking in women in the early twentieth century.

Genetic

Daughters of women with postmenopausal osteoporosis are more likely to develop the condition. Blacks have a lower incidence which may be due to the larger dimensions of their bones and muscles. The larger muscle bulk may confer a protective stress effect.

Other conditions associated with osteoporosis (secondary osteoporosis)

Corticosteroids

Cushing noted the association with osteoporosis when he described the disease of adrenal hyperplasia. In Cushing's syndrome the incidence of osteoporosis is about 50% and that of fracture, 20%. Fractures tend to occur in trabecular bone, particularly the vertebrae. In association with corticosteroid medication the prevalence of fracture may be as much as 17%, depending on the dose (Reid, 1990). As stated above the pathological effect is of trabecular thinning. A number of mechanisms appear to combine stimulating parathyroid hormone, increasing osteoclast activity and accelerating bone resorption. By means of indirect enhancement of resorption and direct inhibition of bone formation, 'uncoupling' is precipitated.

Following commencement of steroids, there is a period of rapid bone loss which lasts for a few months and subsequently tails off. There are no firm conclusions in the literature but dosage over 5 mg a day may precipitate osteoporotic change. There is no evidence that alternate day administration of corticosteroid reduces incidence. No effective combination therapy which can be taken at the same time as steroids to counteract bone loss has yet been introduced.

For the treatment of established osteoporosis secondary to corticosteroids calcium and 25-hydroxy vitamin D supplements, salmon calcitonin and biphosphonates have been shown to reverse the osteoporotic trend by decreasing bone resorption. Drugs which increase bone formation such as sodium fluoride and anabolic steroids increase trabecular bone volume and bone mineral content respectively, although the effect on bone strength is unclear.

Idiopathic juvenile osteoporosis

This curious condition presents with failure to grow, long bone fractures and kyphosis. The histological appearances are of cessation of osteoblastic activity. The serum 1.25 DHCC is commonly reduced. The disease usually responds to calcitriol administration.

Osteogenesis imperfecta (types I–IV)

The various types of brittle bone disease are related to mutations in genes coding for type one collagen alpha chains. The dominantly inherited type I disease closely resembles osteoporosis. In these patients there is a 50% reduction of collagen production thus causing lower bone density and reduced peak bone mass leading to an increased fracture rate.

Conclusion

Osteoporosis is a major underlying factor in the aetiology of fractures in the elderly. These fractures cause enormous morbidity and strain upon health care resources. A clearer understanding of the multifactorial mechanisms underlying the disease will increase the effectiveness of preventive measures. Unfortunately, control of osteoporosis is only likely to have a partial effect on reduction of fracture incidence. Care will also have to be taken in observing the side effects of prophylactic medication.

References

Ansell, B.M. and Lawrence, J.S. (1966) Fluoridation and rheumatic diseases: a comparison of rheumatism in Watford and Leigh. *Annals of the Rheumatic Diseases*, **25**, 67–75.

Compston, J.E. (1990) Structural mechanisms of trabecular bone loss. *Osteoporosis 1990*. (Smith, R. ed.). Royal College of Physicians of London.

Griffin, J. (1990) *Osteoporosis and the Risk of Fracture*. Office of Health Economics, London.

Grimley Evans, J. (1990a) The significance of osteoporosis. In: *Osteoporosis 1990*. (Smith, R. ed.). Royal College of Physicians of London.

Grimley Evans, J. (1990b) Epidemiology: mini-symposium: fractured neck of femur. *Current Orthopaedics,* **4,** 150–55.

Holbrook, T.L., Barrett-Connor, E. and Wingard, D.L. (1988) Dietary calcium and risk of hip fracture: 14-year prospective population study. *Lancet,* ii, 1046–49.

Kanders, B., Dempster, D.W. and Lindsay, R. (1988). Interaction of calcium nutrition and physical activity on bone mass in young women. *Journal of Bone and Mineral Research,* **3,** 145–59.

Kanis, J.A., Aaron, J., Thavarajah, M. *et al.* (1990) In: Osteoporosis: causes and therapeutic implications. *Osteoporosis 1990,* (Smith, R. ed.). Royal College of Physicians of London.

Matkovic, V., Kostial, K., Simonovic, I., Buzina, R., Brodarec, A. and Nordin, B.E.C. (1979) Bone status and fracture rates in two regions of Yugoslavia. *American Journal of Clinical Nutrition,* **32,** 540–49.

Orwoll, E.S., Ferar, J., Oviatt, S.K., Huntington, K. and McLung, M.R. (1987) Swimming exercise and bone mass. In: *Osteoporosis 1987,* (Christiansen, C., Johansen, J.S. and Riis, B.J., eds.). Viborg, Norhaven: pp. 494–98.

Reid, D.M. (1990) Corticosteroid-induced osteoporosis. *Osteoporosis 1990,* (Smith, R. ed.). Royal College of Physicians of London.

Riggs, B.L. and Melton L.J. (1983) Evidence for two distinct syndromes of involutional osteoporosis. *American Journal of Medicine,* **75:** 899–901.

Riggs, B.L. and Melton, L.J. (1986) Ivolutional osteoporosis. *New England Journal of Medicine,* **314,** 1676–86.

Riis, B., Thomsen, K. and Christiansen, C. (1987). Does calcium supplementation prevent postmenopausal bone loss? *New England Journal of Medicine,* **316,** 173–77.

Royal College of Physicians of London (1989) *Fractured Neck of Femur – Prevention and Management.* Royal College of Physicians of London.

Russell, R.G.G. (1990) Bone cell biology: the role of cytokines and other mediators. In: *Osteoporosis 1990.* (Smith, R. ed.). Royal College of Physicians of London.

Singh, M., Nagrath, A.R. and Maini, P.S. (1970) Changes in trabecular pattern of the upper end of the femur as an index of osteoporosis. *Journal of Bone and Joint Surgery,* **52A,** 457–67.

Smith, R. (1987) Osteoporosis: cause and management. *British Medical Journal,* **294,** 329–32.

Stevenson, J.C., Lees, B., Devonport, M., Cust, M.P. and Ganger, K.F. (1989) Determinants of bone density in normal women: risk factors for future osteoporosis? *British Medical Journal,* **298,** 924–28.

Chapter 10

Prevention of infection in orthopaedic surgery

George B. Irvine

While infection poses a threat after any open surgical procedure it has become clear that operations involving the implantation of foreign material are at particular risk. Careful surgical technique coupled with rigorous theatre discipline and the employment of the preventative measures to be discussed should maintain the deep infection rate following implant surgery at a level below 1%.

Introduction

The practice of antiseptic techniques followed the pioneering work of Lister during the last century. Although he recognized contamination from the surrounding air, his dramatic success in reducing sepsis rates and mortality was mainly attributed to sterilization of the instruments and operative field and his carbolic aerosol spray rapidly fell into disuse.

During this century the discovery of penicillin was a conspicuous landmark and the subsequent development of antimicrobial chemotherapy has made the treatment of specific infections possible. However, in spite of further reductions in gross sepsis and mortality postoperative infection has remained a persistent problem.

More recently the routes and sources of infection in the operating theatre have been established as a result of epidemiological research and this has led to the control of air movement and an appraisal of ventilation methods. Plenum systems have become standard and, while they maintain the operating theatre at a positive pressure with respect to surrounding areas, they give rise to a highly turbulent pattern of air flow which can lead to the circulation of dust and other particulate debris. Laminar flow systems direct highly filtered air into the operative field but remain capable of causing turbulence and entrainment of particles at the periphery of the air column unless the unit is walled-off or the velocity of the air flow is graduated in exponential fashion from the centre of the column. Ultraviolet irradiation has a wavelength-specific bactericidal effect and while it is widely used in North America it has never become popular in the UK.

Molecular mechanisms in sepsis

Scrutiny at a molecular level is necessary in order to understand the susceptibility of implants to infection and, in this context, it appears that tissue integration may be crucial in preventing bacterial infection.

Infection around biomaterials involves adhesion to, and colonization of, their surfaces. Tissue integration and bacterial adhesion are chemically parallel and competitive processes but appear to be mutually exclusive. Surfaces colonized by healthy tissue cells with intact membranes are resistant to infection, whereas bacterial colonies on biomaterial surfaces are seldom replaced by tissue cells and are resistant to both treatment and host defence mechanisms.

Implanted biomaterials provide a favourable environment for bacterial colonization because they provide surfaces that facilitate chemical, physical and biological interaction. Bacteria may form an attachment using specific receptor-ligand chemical bonds or non-specifically by charge-related, hydrophobic and extracellular polysaccharide-based interactions. Proteinaceous adhesins, polysaccharide polymers and other substances in the milieu are then able to interact to form a 'slime' or biofilm. This confers resistance to host defence mechanisms such as phagocyte engulfment, surfactants and antibodies as well as resistance to antibiotics. After implantation biomaterial surfaces are disturbed by corrosion, wear and chemical interactions which result in the release of ions or molecules into the peri-implant matrix. These changes may both enhance bacterial virulence as well as diminishing macrophage effectiveness.

Ultraclean air

Sir John Charnley's initial success with total hip replacement in the early 1960s was associated with a high rate of sepsis, often approaching 10%, and presenting at a late stage. He believed that air-borne contamination was the principal causative factor and took steps to introduce and gradually refine a system of special clothing and ventilation. With these measures he was able to show a tenfold reduction in his infection rate. Although Charnley's argument in favour of clean air was compelling, other groups had achieved similar results with the use of prophylactic antibiotics alone and, for this reason, in 1973 The Medical Research Council agreed to carry out a fully controlled investigation into the effect of reducing air contamination in the operating theatre on the incidence of sepsis after total joint replacement operations.

Statistical considerations demanded a large scale study and ultimately over 8000 primary hip and knee replacement operations were included from 19 hospitals in the UK and Sweden. Each participating surgeon operated in both a conventionally ventilated theatre in which staff wore ordinary cotton clothing and an ultraclean air theatre with either ordinary or special clothing. Patients were allocated to the environment by a randomizing procedure and there were no differences in procedures between the groups. Air samples were taken in all the operating theatres during surgery and a proportion of wounds were sampled bacteriologically.

The main conclusions of the study are shown in Table 10.1. The level of air-borne contamination was consistently reduced in the ultraclean air theatres although, even with similar ventilation systems, there was a wide variation between individual hospitals. This may be attributed, at least in part, to the known variation in bacterial dispersion between individuals. Clothing designed to reduce bacterial dispersion was effective whatever the method of ventilation but its effect was most marked in ultraclean air theatres. Analysis of the wound washout data suggested that the great majority of wound contamination in the control group was derived from the air.

Table 10.1 Main findings at the MRC trial

Incidence of deep infection mirrored level of air contamination
Use of ultraclean air system reduced infection rate by one-half
Addition of body exhaust suits reduced infection rate by a further one-quarter
Prophylactic antibiotics were beneficial in all groups

Follow up extended to one year after the last operation recorded at a given hospital (i.e. one to 4-year follow up for the whole trial population). The incidence of re-operation in the absence of sepsis was not significantly different between the groups but the incidence of infection was much less frequent after operations in a clean air theatre (Table 10.2). The protective effect of special clothing was proven. *Staphylococcus aureus* and *Staphylococcus epidermidis* were the commonest bacterial species in 68 patients in whom presumptively causative organisms were isolated. The risk of subsequent deep infection was shown to be substantially increased if there had been any suspicion of postoperative wound sepsis. Such superficial wound infection was found to be most often due to organisms acquired outside the operating theatre since ultraclean air had little effect on its incidence.

Table 10.2 Rates of deep infection according to ventilation, theatre clothing and use of antibiotics. MRC trial

		Incidence of deep infection (%)	
Ventilation system	*Theatre clothing*	*No antibiotics*	*Prophylactic antibiotics*
Control (turbulent)	*Conventional*	*3.4*	*0.8*
Ultraclean	*Conventional*	*1.6*	*0.7*
Ultraclean	*Body exhaust*	*0.9*	*0.06*

Extended follow up to a median time of 7 years was applied to a cohort of 85 patients who had suspected deep infection but in whom re-operation had not been performed at the end of the initial study period. These patients fared badly and infection was subsequently confirmed in approximately one-quarter. Of equal interest, however, is the fact that four of the matched controls developed deep infection during this period of extended follow up. This reflects late presentation and suggests that the figures from the MRC trial may underestimate by one-half the number of patients likely to develop sepsis in the 7 years following primary operation.

Theatre clothing

Clothing is of paramount importance since it is the personnel who are the principal source of bacteria in the operating theatre. The routes of transfer of these bacteria are by air and contact, and the concentration of bacteria relates to both the number of people in the operating theatre as well as their activity. Clearly it is the surgical team rather than the patient who are responsible for the bulk of the air-borne microbial contamination. Bacteria may issue from the open apertures of clothing as well as passing directly through the fabric by diffusion or capillary action.

Standard cotton clothing and gowns are both comfortable and cool to wear. These attributes result from the loose weave which gives an average pore size of 80 μm. Since bacteria carrying particles average 12-14 μm the filtering efficiency of cotton is poor. Special treatment is required for water proofing and this is likely to be removed by laundering. As a natural fabric it is prone to shrinkage and surface deterioration or 'linting'. Manufacturing techniques may be varied in order to produce a tighter weave as in the 'Ventile' cloth used for the Charnley body exhaust suit. This system is highly effective in reducing air-borne bacterial contamination, but it is cumbersome in use and has never become popular. Furthermore, developments in textile technology have occured so that conventional clothing made from alternative fabrics may be as effective when used in conjunction with a vertical laminar flow ventilation system.

Disposable fabrics are usually paper-based. They are non-woven and allow easy passage of air. They derive their filtering efficiency from the fact the bacteria carrying particles must twist and turn through the random matt of fibres. Breathable membrane fabrics are occlusive but washable and the best known is 'Gore-tex'. This is a fabric in which a layer of PTFE film is laminated to one or two layers of a woven fabric such as polyester. The plastic film has a pore diameter of 0.2 μm which makes it a good bacterial filter but hinders air exchange. The fabric is further likely to deteriorate with laundering so that special arrangements are necessary.

Woven polyester fabrics appear to be the most promising new textiles. They have resulted from clean room technology in the micro-electronics and pharmaceutical industries. These fabrics are extruded polymers and can be manufactured so that they are both protective and comfortable. They can provide an efficient liquid barrier and excellent bacterial filtering capacity while retaining good air exchange. They are non-linting, antistatic, can be sterilized by standard processes and undergo minimal surface deterioration with time (Figures 10.1, 10.2 and 10.3).

Figure 10.1 Photomicrographs of (a) cotton and (b) polyester (Rotecno) fabric prior to use

Figure 10.2 Photomicrographs of (a) cotton and (b) polyester (Rotecno) fabric after 1 and 20 months respectively

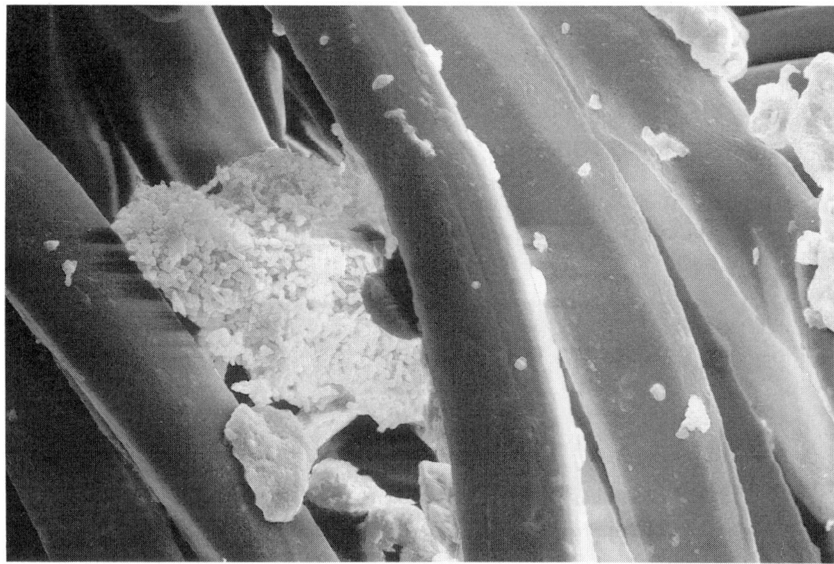

Figure 10.3 Photomicrograph of a bacteria-carrying particle trapped on the surface of polyester (Rotecno) fabric

Prophylactic antibiotics

The results of the MRC Trial clearly demonstrated the benefit of antibiotic prophylaxis in joint replacement surgery irrespective of the type of clothing or ventilation method. Activity against *Staphylococcus aureus* and *Staphylococcus epidermidis,* coliforms and anaerobes is required (Table 10.3) and this can only be achieved by combination therapy. The agents used should be bactericidal, potent, relatively free from side effects and possess a wide therapeutic index. Cost is an additional consideration.

Table 10.3 Types of organisms in hip implant infections

	Staphylococcus aureus	*Staphylococcus epidermidis*	*Coliforms*	*Anaerobes*	*Others*
Fitzgerald *et al.* (USA, 1977)	10	8	12	10 (7)	9
Nelson (USA, 1977)	5	1	4	4 (3)	1
Andrews *et al.* (UK, 1981)	32	15	36	4	15
Buchholz *et al.* (FDR, 1981)	274	–	145	138 (74)	38
Kamme and Lindberg (Sweden, 1981)	7	9	7	17 (10)	4
Charnley and Amstutz (USA, 1983)	10	7	12	3 (1)	1
Surin *et al.* (Sweden, 1983)	80	28	36	37	

( ) Peptococcus spp.

In constructing an antibiotic regimen, priority should be given to ease of administration and to antibiotics which give satisfactory cover to as many of the different groups of organisms likely to cause infection as possible. The number of agents should also be kept to a minimum to reduce the chance of side effects or interactions and to simplify administration. The ideal combination at the present time appears to be cefuroxime and metronidazole (Table 10.4). Clindamycin and gentamicin also gives excellent cover but this combination is much more capable of producing side effects. Flucloxacillin, gentamicin and metronidazole has no obvious advantage over the two-agent regimens.

Table 10.4 Activity of different antibiotic combinations

	Staphylococcus aureus	*Staphylococcus epidermidis*	*Coliforms*	*Anaerobes*
Cefuroxime/metronidazole	+++	+++	+++	+++
Flucloxacillin/gentamicin	+++	++	+++	+
Flucloxacillin/gentamicin/metronidazole	+++	++	+++	+++
Erythromycin/gentamicin	++	++	+++	++
Clindamycin/gentamicin	+++	++	+++	+++
Augmentin/gentamicin	++	++	++	+

With regard to timing of administration, it is probable that the best levels of antibiotics in the tissues are achieved by giving the agents intravenously at the start of the operation. In terms of maintaining adequate blood levels during surgery, the half-life of each agent must be considered. This is generally about 2–3 hours for the cephalosporins. Consequently in a prolonged operation and in operations where there may have been substantial blood loss and replacement it is appropriate to replenish antibiotic levels by a repeat dose during the operation. As a general rule half the original dose should be given 2 hours after the start of surgery for the regimens described. It is far more important to have sufficient levels of antibiotic at all times during the operation than to give lower doses for prolonged periods beyond the operation. Current knowledge supports the use of two postoperative doses and certainly prolonged prophylaxis is to be avoided after routine primary operations.

Acknowledgements

Tables reproduced from *Journal of Hospital Infection* with permission. Photographs kindly supplied by Lojigma International Limited.

References

Charnley, J. (1979) *Low Friction Arthroplasty of the Hip*. Berlin, Heidelberg, New York, Springer Verlag.

Lidwell, O.M., Lowbury, E.J.L., Whyte, W., Blowers, R., Stanley, S.J. and Lowe,

D. (1982) Effect of ultraclean air in operating rooms on deep sepsis in the joint after total hip or knee replacement; a randomised study. *British Medical Journal*, **285:** 10–14.

Lidwell, O.M. (1988) Air, antibiotics and sepsis in replacement joints. *Journal of Hospital Infection*, **11** (suppl. c), 18–40.

Molecular mechanisms in musculo-skeletal sepsis: the race for the surface (1990). *Instructional Course Lectures* 1990; **39:** 471–482.

Sanderson, P.J. (1988) The choice between prophylactic agents for orthopaedic surgery. *Journal of Hospital Infection*, **11** (suppl. c), 57–67.

Whyte, W. (1988) The role of clothing and drapes in the operating room. *Journal of Hospital Infection*, **11** (suppl. c), 2–17.

Chapter 11

Scientific basis of MRI scanning and its use in orthopaedics

Roy S. Twyman

The phenomenon of magnetic resonance (MR) was discovered by Bloch and Purcell in 1947, and it has been used extensively for chemical analysis. However, it was 1973 before the first MR image was published by Lauterbur working in New York. Since then there has been a rapid evolution in techniques such that magnetic resonance imaging (MRI) now produces extraordinary images of diagnostic potential.

Magnetic resonance is defined as the enhanced absorption of energy occurring when the nuclei of atoms or molecules within an external magnetic field are exposed to radiofrequency energy at a specific frequency called the Larmor frequency.

Before discussing the clinical aspects, the definition and some of the basic physics of how the images are formed will be explained.

Physical principles

The body is relatively transparent to X-rays at one end of the electromagnetic spectrum and to radiowaves at the other while remaining opaque to intermediate wavelengths. It is only within these transparent areas that tissue can be imaged.

In contrast to X-rays, radiowaves are non-ionizing, very low energy radiation.

Chemical properties are generally related to electron structure of atoms, but the physical properties are more closely related to the nucleus.

Any nucleus with an odd number of protons or neutrons can produce a MR signal. The ideal atom for medical imaging is hydrogen whose nucleus is made up of one unpaired proton and one unpaired neutron making it easy to manipulate in a magnetic field. It also is in high abundance forming two thirds of the atoms in the human body. The signal generated will be large allowing for high definition images.

Protons spin about their axes and so have angular momentum. They also carry a charge and therefore, according to Faraday's law of induction, produce a magnetic field acting like tiny bar magnets.

Normally the vectors of the proton spin axis, and so their magentic fields, are at random (Fig. 11.1), but when placed in a static, or constant, magnetic field they line up with it (Fig. 11.2). Unlike larger magnets which obey

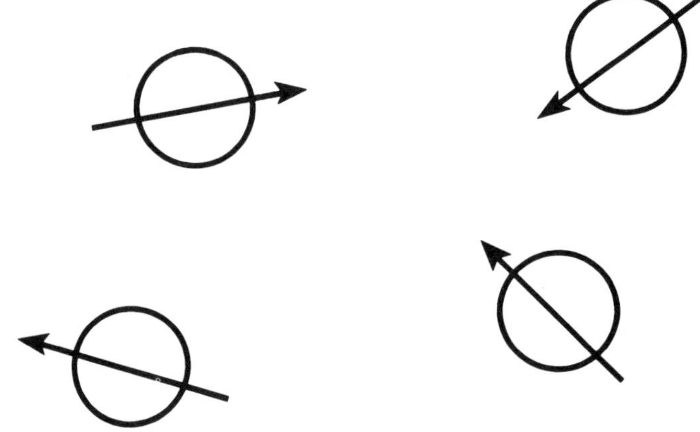

Figure 11.1 Randomly oriented protons

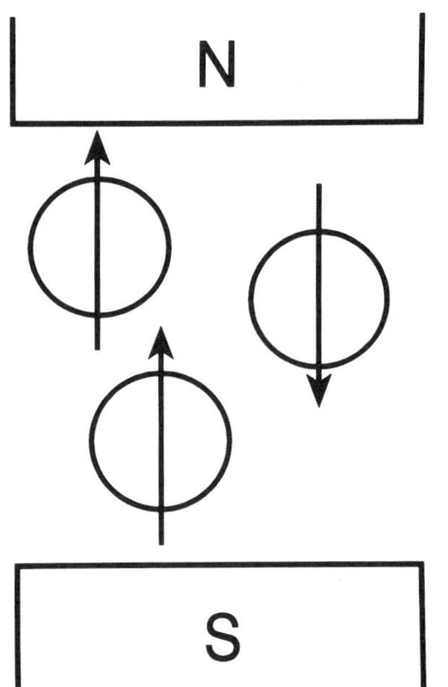

Figure 11.2 Protons lining up parallel or anti-parallel in a magnetic field

classical Newtonian physics, the rules of quantum physics allow them to be either in line (parallel), or at 180° (anti-parallel) to the magnetic field. These are low and high energy states respectively.

In any magnetic field there is a net excess of nuclei in the low energy state. This is the net magnetization M and it produces the MR signal, M and hence the signal strength increases in direct proportion to the external magnetic field strength.

The spinning protons are not exactly in line with the field but wobble around it in much the same way as a gyroscope under the influence of gravity. This movement is known as precession (Fig. 11.3). The phases of precession of the spin in different protons are at random, i.e. some are at one o'clock, some at five o'clock and so on. Therefore the net magnetic moment of all the protons, known as M, is in line with the static field.

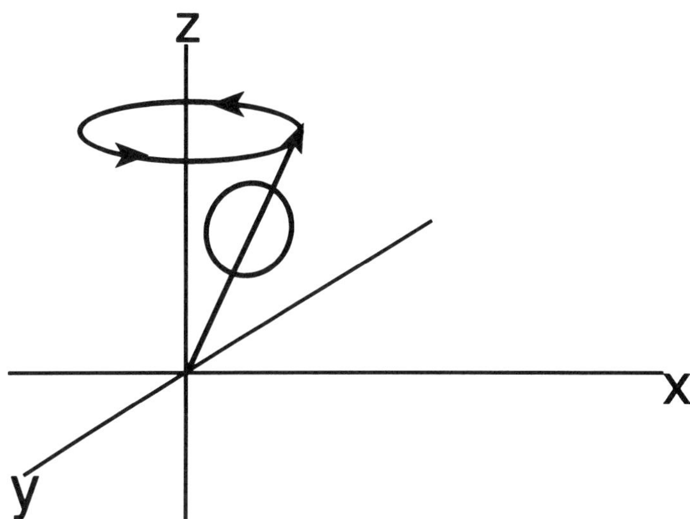

Figure 11.3 Precession

Larmor's relationship states the rate of precession or angular frequency is proportional to field strength. That is the stronger the field the faster the precession or 'wobble'. This is a fundamentally important principle of MRI.

A radiofrequency (RF) pulse is now added at 90° to the static field at the same frequency as the rate of precession. It affects the spins forcing them to precess at an ever wider angle up to 90° to the static field. The phases of the precession will also converge so the axes of all the spins will be pointing in the same direction at any one time. The protons will then be resonating with the radiofrequency and there will be a rotating net magnetic field in the transverse or xy plane.

The angle by which the RF pulse rotates M into the xy plane is called the flip angle. Any angle can be applied depending on the image required.

Faraday's law of induction states that movement of a magnetic field induces a voltage in receiver coils. With the magnetic field rotating in the transverse plane, the amplitude of the induced voltage or signal is proportional to the number of spins at that location, i.e. the proton density. Its frequency will be at the Larmor frequency, i.e. at the rate of precession.

The radiofrequency is then removed leaving the spins solely under the influence of the static field. The magnetic moment M will then 'free precess' and gradually return to the rest state. The induced voltage in the receiver coils will diminish as it is only the vector of M which remains in the transverse plane inducing a voltage. This is known as the free induction decay or FID.

Relaxation times

Up till now the signal discussed has only built up a map of proton density. In order to learn about the locations and their environment the relaxation times or the time taken to return to the rest state must be studied.

The diminution of the signal is studied in two planes giving T_1 and T_2 relaxation times.

T_1 or the spin lattice relaxation time is the time taken to restore the net magnetic moment M parallel to the static field. It depends on the lattice structure in which protons are held. Protons held tightly within crystals have a long T_1, whereas protons in water are free to return to the rest state rapidly and so have a short T_1. Tissues with a short T_1 will appear bright.

Various sequences of RF pulses are used to alter the image contrast. The inversion recovery sequence is very T_1 weighted. It begins with a 180° pulse instead of the 90° pulse so the spins relax over twice the range, essentially doubling the contrast.

The predicted drop off in the induced voltage is very much slower than that which happens in reality. While the RF pulse is on all the spins are in phase and so the net magnetic moment M is a single line, but when it is removed M gradually fans out with some protons spinning faster than others (Fig. 11.4). After a while the faster ones will be 180° out of phase with the slower ones and so cancel each other out. The induced voltage is thus reduced.

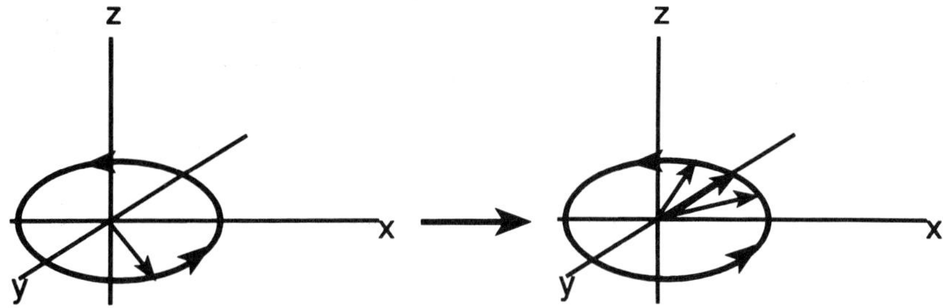

Figure 11.4 Loss of synchronous phase with time

The reasons for this are twofold. In practice it is not possible to produce a static field which is completely homogeneous and protons are under slightly different magnetic field strengths, hence precess at different rates. The second factor is that spinning protons interact with each other's magnetic field. Therefore the rate of a particular proton's spin will depend on its local environment.

T_2 or the spin-spin relaxation time is the time taken for this loss of phase coherence in a perfect field.

The spin echo sequence has been devised in order to remove the influence of inhomogeneity of the field. If during the decay a 180° RF pulse is given, the spins will flip to a mirror image, i.e. the fast ones will be behind the slow ones. They will then continue to rotate and reform a single line, M, once again at the starting point as the fast spins catch up with the slow ones. The signal will fade as the field, M, fans out once again. The 180° pulse is repeated several times. Each echo peak will be smaller than its predecessor due solely to spin interaction. This will be independent of the field inhomogeneity and so a decay in the signal will be purely dependent on the nuclear interactions.

The time between the initial RF pulse and the peak of the echo is the echo time (TE).

Image reconstruction

In order to know where a signal is coming from each point in the body must be given a unique signal or frequency. The signal must then be encoded to give spatial information. It is then digitized and decoded using a mathematical process known as the Fourier transform. This is able to extract the individual frequencies and their amplitude from a complex signal. It is analogous to one's ear picking out a particular instrument in an orchestra.

If a linear magnetic gradient along the length of the body is superimposed on the static magnetic field along the z axis at the same time as the initial RF pulse (Fig. 11.5), then different locations will have different field strengths and according to the Larmor relationship the protons will precess at different rates. If the RF is of a single frequency only a narrow slice of precessing protons will be of that precise frequency and so resonant. A single slice in the transverse plane of the body has been isolated.

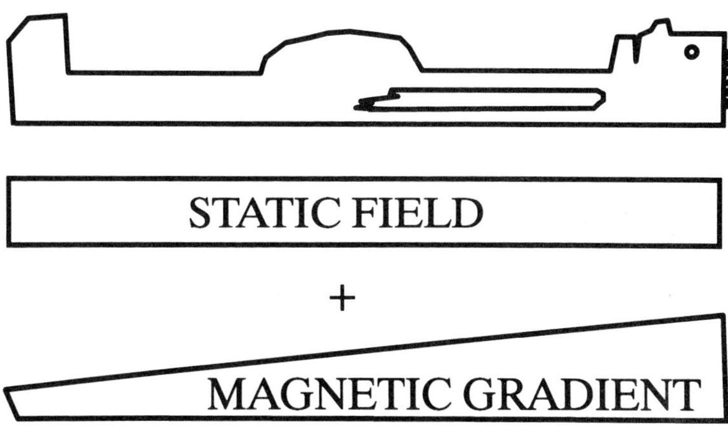

Figure 11.5 Addition of a magnetic gradient to the static field

Other gradients in the x and y planes are used to define where, within the slice, the signal is coming from, so producing a map or image.

Three-dimensional images are built up using a more complex form of the Fourier transform. In this type of image formation the whole object is excited simultaneously. It is known as volume imaging and it is obviously much quicker than collecting information from slices. Computer software is then able to form an image in any plane from these data.

The image can be improved by placing receiver coils in close proximity to the area of interest thereby increasing the signal to noise ratio. Surface coils are used against the spine enhancing the image of the posterior structures, but excluding the motion artefacts derived more anteriorly. Coupled coils surround large joints such as the knee or shoulder.

Safety

There are few contraindications to MR scanning. Patients with pacemakers and women in the first trimester of pregnancy are excluded. Some aneurysm clips are ferromagnetic and so may be dislodged, with severe consequences. A small proportion of patients become claustrophobic within the scanner and so are unable to remain within it.

Orthopaedic implants rarely cause a problem. The image is distorted less than on CT and is confined to the immediate vicinity around the metal. There is a theoretical risk of the RF inducing eddy currents in the prosthesis causing heating. In vitro studies have shown this not to be significant.

MRI in orthopaedics

MRI is non-invasive and so is preferable to arthrography or arthroscopy of the knee. It is excellent at staging both bone and soft tissue tumours and demonstrating involvement of adjoining muscles, vessels and nerves. It is useful in demonstrating traumatic conditions of soft tissue. It has an important role in evaluating degenerative and neoplastic conditions of the spine. Its major advantage over CT is in the excellent depiction of soft tissue contrast that can be manipulated by varying the sequence parameters. There is also a multiplanar facility. All this is done without irradiating the patient.

In order to make sense of the images MR produces one must know the signal characteristics of various tissues (Table 11.1). Tissues with low proton densities and high collagen content have a low signal, i.e. dark on all MR images. The signal from cancellous bone varies with its contents.

Table 11.1 Signal Characteristics of various tissues

High signal on both T_1 and T_2 weighted images	Fat, haematoma, slowly flowing blood
Low signal on both T_1 and T_2 weighted images	Cortical bone, ligaments, tendons, fast flowing blood
Intermediate signal on T_1 and T_2 weighted images	Muscle, nerves, hyaline cartilage
Low signal on T_1 but high intensity on T_2	Haematopoietic marrow, CSF, nucleus pulposus, most tumours, infection, cysts

Tumours

The excellent contrast resolution and multiplanar imaging of MRI allows for much greater definition than with plain films or CT of the extent of soft tissue and bone tumours. Soft tissue tumours are poorly defined by CT and angiography is often required as an adjunct, but, on MRI of malignant tumours, there is excellent delineation from surrounding muscles. The intramedullary extent of bone tumours is best assessed on MRI (Fig. 11.6),

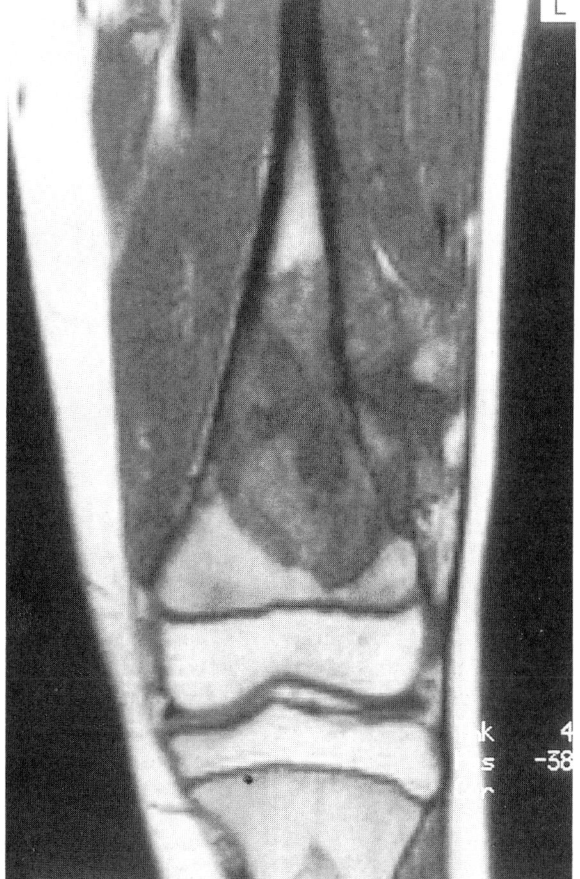

Figure 11.6 T_1 weighted image of a distal femoral osteosarcoma

as is staging the spread into soft tissues, though oedema and tumour may be impossible to tell apart. Cortical bone destruction, periosteal reaction, and calcification remain superior on conventional radiology.

Unfortunately, the specificity of tumours on MRI is poor. There are no specific characteristics which distinguish benign from malignant, the signals on T_1 and T_2 images and the heterogeneity of the lesion can be similar. Fibrous tumours appear dark and those with high fat content are bright, but the exact pathology cannot, at present, be given.

The information MRI can provide in assessing the extent of the tumour is helpful in surgical planning with a growing trend for reconstructive procedures, i.e. the extent of bone and soft tissue involvement, which compartments are affected and the involvement of the neurovascular bundle. The presence of skip lesions can be seen.

The signal intensity changes following a response to either radiotherapy or chemotherapy, so it is hoped their effectiveness can be assessed with sequential scans (Fig. 11.7). The difficulties arise in differentiation of radiation change, fibrosis and recurrent tumour which can be indistingishable.

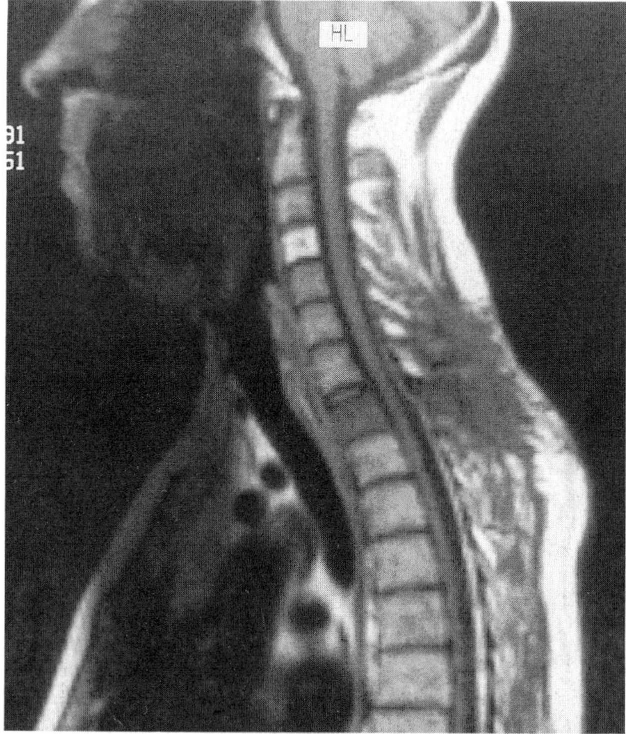

Figure 11.7 T_1 weighted image of a myeloma affecting T1. C4 has previously been irradiated

Spine

MRI most clearly shows its clinical usefulness in the assessment of spinal pathology. The condition of the vertebral body, disc, nerve roots and theca are all well shown with the aid of surface coils.

The assessment of cervical and lumbar radiculopathy is as accurate as CT myelography. The nerve root signal is relatively low in intensity and is outlined by fat with its high signal. The normal disc has a low signal annulus and a higher signal from the nucleus on T_2 weighted scans. With degeneration, as the water content decreases, this difference becomes less distinct. These changes correlate well with discography. Herniation of the disc is identified by a focal impression on the anterior epidural fat. Root compression and displacement are readily seen using a combination of axial and sagittal images.

Several discs are seen on the same scan. This has proved particularly useful before spinal fusion to decide which levels are amenable to surgery without resorting to discography. Five to six levels can be imaged on one scan with the aid of surface coils.

Postoperative scarring and arachnoiditis can be distinguished from recurrent disc herniation with the aid of the paramagnetic contrast agent gadolinium-DTPA (Gd-DTPA). Fibrosis shows enhancement whereas disc prolapse does not (Fig. 11.8).

Deformity or displacement of the spinal cord is also demonstrated, for

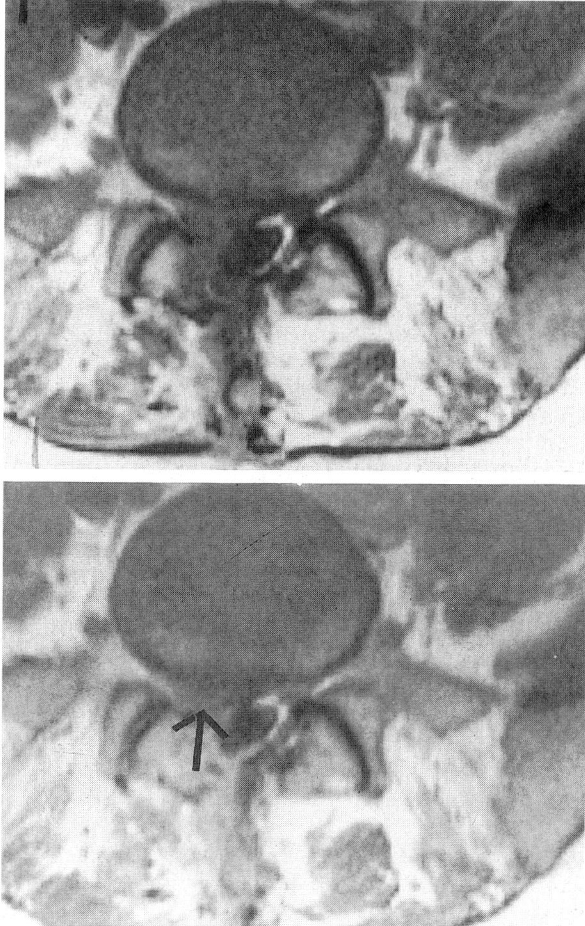

Figure 11.8 T_1 weighted image (a) pre- and (b) post- i.v. gadolinium of the L4/L5 disc. The patient had recently had a discectomy. The postoperative fibrosis enhances but the recurrent disc prolapse (arrowed) does not

example at the atlantoaxial joint in rheumatoid disease. Flexion and extension scans can be performed. Spinal compression following vertebral fractures can be seen. Intracordal injury is also better assessed than by any other means.

The cavitation and extent of syringomyelia is readily assessed, though differentiation from cystic tumours may be difficult. This may also be helped by Gd-DTPA. More information can be gained about spinal tumours than either by CT or myelography. In particular, the extent into the cord and its location can be fully evaluated.

Disc space infection is seen earlier by MRI than any other means of investigation. A diffuse increase in the signal on T_2 images of the whole disc and adjacent end plates is observed. Extension into the extradural space can be picked up unlike scintigraphy.

Knee

The menisci are readily assessed on coronal and sagittal slices with the aid of surface coils which increase the spatial resolution and the signal to noise ratio. It is now at least as accurate as arthrography in the investigation of tears and if widely available could replace both arthrography and arthroscopy as a diagnostic means (Crues *et al.*, 1987). Like arthrography the posterior horn is easily seen. The meniscus is normally of a low signal. A tear is shown by an abnormally high linear signal, probably representing synovial fluid, extending to the articular surface (Fig. 11.9). If the signal does not extend the whole way to the surface it is likely the meniscus is undergoing myxoid degeneration.

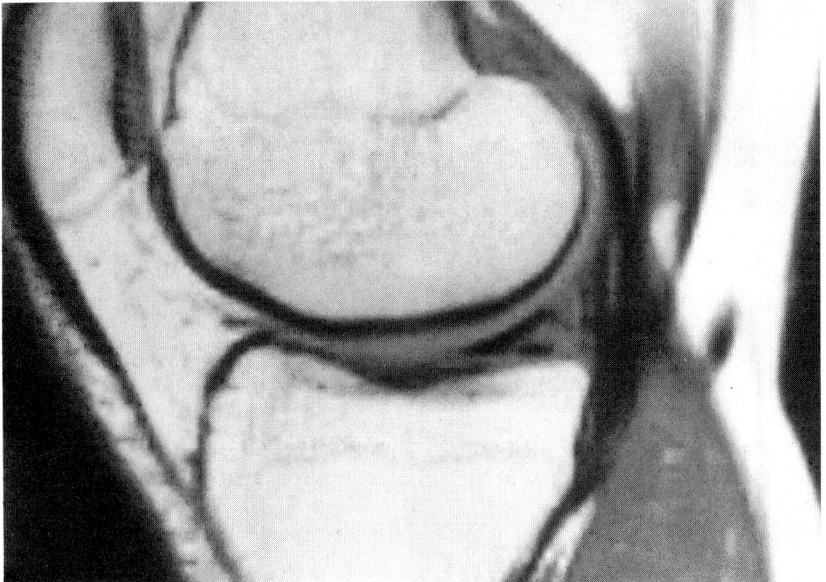

Figure 11.9 T_1 weighted image of a tear of the posterior horn of the medial meniscus

All the ligaments of the knee are well seen. Complete tears of the cruciates are easily identified, but there are difficulties with partial lesions.

Articular cartilage is well seen. This is likely to be helpful in erosive arthropathies. The state of the overlying cartilage in osteochondritis dissecans can be assessed thus aiding the planning of treatment. Effusions are readily seen and distinguished from synovial thickening.

Hip

Most interest in the hip has centred around the early detection of avascular necrosis of the femoral head. This diagnosis can be made earlier on MRI than on any other means and therefore treatment can be instituted before mechanical failure of bone has occurred. The distinctive feature is the reactive interface between normal and necrotic bone (Fig. 11.10). This can

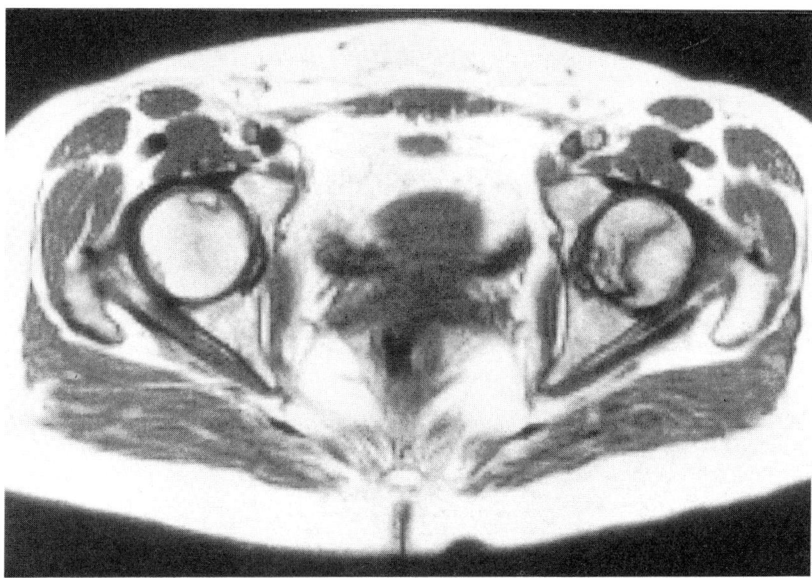

Figure 11.10 T_1 weighted image of bilateral avascular necrosis of the femoral head, much more advanced on the right

be seen on both T_1 and T_2 sequences, but a specific double line can usually be seen on T_2 weighted images. The T_1 images in advanced disease occasionally cannot be distinguished from tumours. MRI maybe useful in selecting patients for treatment and sequential scans are able to monitor progress.

Perthes disease is well demonstrated. MRI is at least as sensitive as scintigraphy in early disease. Long-term follow up without the need for multiple X-rays in children is possible.

The exact anatomy of congenital dislocation of the hip can be displayed and so the management can be more easily decided.

Shoulder

Initial difficulties due to the shoulder being at the periphery of the magnetic field have been overcome with surface coils. The rotator cuff can be well visualized. Full and partial tears are identifed thus giving it a distinct advantage over arthrography which it is likely it will replace as more MRI scanners become available. The site of impingement, offending osteophytes, the size of the tear and retraction of tendons can be defined and surgery planned.

In instability of the shoulder, labral tears are seen as bright signals within the normally dark fibrocartilage. Defects in the humeral head are also well seen.

References and further reading

Berns, D.H., Blaser, S.I. and Modic, M.T. (1989) Magnetic resonance imaging of the spine. *Clinical Orthopaedics*, **244**, 78–100.

Crues, J.V., Mink, J., Levy, T.L. *et al.* (1987) Meniscal tears of the knee: accuracy of MR imaging. *Radiology*, **164**, 445–448.

General Electric Company. NMR – A perspective on imaging.

Lauterbur, P.C. (1973) Image formation by induced local interactions: Examples employing nuclear magnetic resonance *Nature*, **242**, 190.

Mink, J.H. and Deutch, A.L. (1989) Occult cartilage and bone injuries of the knee: detection, classification and assessment with MR imaging. *Radiology*, **170**, 823–829.

Mitchell, D.G., Kressel, H.Y., Arger, P.H. *et al.* (1986) Avascular necrosis of the hip: comparison of MR, CT and scintigraphy. *American Journal of Radiology*, **147**, 67–71.

Modic, M.T., Feiglin, D.H., Piraino, D.W. *et al.* (1985) Vertebral osteomyelitis: assessment using MR. *Radiology*, **157**, 157–166.

Modic, M.T., Masaryk, T., Boumphrey, F. *et al.* (1986) Lumbar herniated disc disease and canal stenosis: prospective evaluation by surface coil, CT and myelography. *American Journal of Radiology*, **147**, 757–765.

Pettersson, H., Gillespy, T., Charters, J.R. *et al.*(1987) Primary musculoskeletal tumours: examination with MR imaging compared with conventional modalities. *Radiology*, **164**, 125–133.

Philips Medical Systems. Principles of MR imaging.

Reicher, M.A., Hartzman, S., Bassett, L.W. *et al.* (1987) MR Imaging of the Knee. *Radiology*, **162**, 547–551.

Ross, J.S., Delamarter, R. and Hueftle, M.G. *et al.* (1989) Gadolinium-DTPA enhanced MR imaging of the postoperative lumbar spine: time course and mechanism of enhancement. *American Journal of Radiology*, **152**, 825–834.

Seeger, L. (1989) Physical principles of magnetic resonance imaging. *Clinical Orthopaedics*, **244**, 7–16.

Seeger, L.L. (1989) Magnetic resonance imaging of the shoulder. *Clinical Orthopaedics*, **244**, 48–59.

Sundaram, M., McGuire, M.H., Herbold, D.R. *et al.* (1986) Magnetic resonance imaging in planning limb salvage for primary malignant tumours of bone. *Journal of Bone and Joint Surgery*, **68A**, 809–819.

Tyrrell, R.L., Glukert, K., Pathria, M. and Modic, M.T. (1988) Fast three dimensional MR imaging of the knee: comparison with arthroscopy. *Radiology*, **166**, 865–872.

Worthington, B.S. (1990) Magnetic resonance of the musculoskelatal system. *The Radiology of Skeletal Disorders*, 3rd edn. Churchill Livingstone, Edinburgh, pp. 1991–2013.

Wuisman, P. and Enneking, W.F. (1990) Prognosis for patients who have osteosarcoma with skip metastasis. *Journal of Bone and Joint Surgery*, **72A**, 60–68.

Index